The User's Guide to Multidimensional Scaling

A vast similitude interlocks all
All distances of space however wide
All distances of time,
All souls, all living bodies though they be
  ever so different,
All nations, all identities that have existed
  or may exist
All lives and deaths, all of the past, present,
  future
This vast similitude spans them, and always
  has spanned,
And shall forever span them and shall compactly
  hold and enclose them.

Walt Whitman, *Leaves of Grass*

The User's Guide to Multidimensional Scaling

with Special Reference to the MDS(X) Library of Computer Programs

A. P. M. Coxon

with the assistance of P. M. Davies

Heinemann Educational Books

Published in the U.S.A. 1982 by
Heinemann Educational Books Inc.
4 Front Street, Exeter, New Hampshire 03833

Library of Congress C.I.P. Data

Coxon, Anthony Peter Macmillan.
  The User's Guide to Multidimensional Scaling.

    Includes index.
  1. Multidimensional Scaling.   2. Multidimensional
Scaling—Computer Programs.
I. Davies, P. M. (Peter M.)   II. Title
H61.C685   1982   001.4'226   82–9317
ISBN 0–435–82251–9   AACR2

Printed in Great Britain

3-30-85

Contents

Preface

This book represents the result of twelve years of research and teaching in the field of multidimensional scaling (MDS) and its applications. During this time the range of MDS had changed remarkably. Initially the only data considered suitable for scaling consisted of subjective estimates of the similarity of stimuli. Subsequently scaling programs have been developed to deal with data and models which are now capable of analysing anything from a contingency table to a free-sorting or repertory grid experiment.

And yet MDS is still considerably under-used. A picture of data is worth a thousand words (as Tukey reminds us) and no preference theorists can afford to be unaware of the relevance of scaling models to their domain. Certainly, social and life scientists are now becoming familiar at least with the basic MDS model and some of its variants. But twelve years is a long time—at least long enough for initial enthusiasm to be tempered and wilder claims to be tested. MDS is neither an automatic machine for turning sows' ears into silk purses nor is it a cruel artifice to make all data submit to a spatial representation at whatever cost.

In this volume I have tried to communicate some of the enthusiasm I still feel for the subject and direct attention to its possible applications. Other books exist either to give a short elementary introduction (Kruskal and Wish's *Multidimensional Scaling* does this admirably) or to give a definitive and exhaustive technical treatment of the area (Forrest Young *magnum opus* (1982) and Schiffman et al. (1982) promise to be such). This book, by contrast, provides the *user* of MDS with a clear outline and exposition of the main model and its chief extensions, and does so with a minimum of technical fuss and by providing a maximum of information. It is chiefly designed for use with the MDS(X) library of programs but is neither dependent on them nor restricted to them. (The development of MDS(X) has been funded in a series of grants from the Social Science Research Council for eight years, for which support we are most grateful.)

The User's Guide forms one of three publications of the MDS(X) Project. A detailed description of the constituent programs of the MDS(X) series and of the command and parameter language for running them is contained in the *MDS(X) User Manual (SV3)* published by the Program Library Unit, University of Edinburgh (18 Buccleuch Place, Edinburgh EH8 9LN) which is distributed with all machine range versions of the MDS(X) package. The *User Manual* is written chiefly by Peter Davies who has acted as documenter during the major part of the Project.

A companion book of readings entitled *Key Texts in Multidimensional Scaling* is published by Heinemann Educational Books and has been produced by Peter Davies and myself to provide a selection of the most definitive and crucial theoretical and applied work in the field: it is intended to provide a resource book for the other two publications.

On a first reading of *The User's Guide*, Chapters 2 and 6 may be omitted without disturbing the flow of the text. Chapter 2 deals entirely with measures of similarity, and whilst it is necessary to understand the content of Chapter 5 before consulting Chapter 6, most users are likely to want to refer directly to the particular program that interests them.

Years of teaching in the area of MDS have convinced me that researchers and students most value a conceptual and applications-oriented exposition where mathematical and technical details are kept to a minimum and obtrude only where it is necessary to make an important substantive or interpretative point. This has been my approach in writing *The User's Guide*. Necessary mathematical (and especially geometrical) background is relegated to technical appendices following the chapters where it becomes most relevant. Readers wanting a more detailed and systematic algebraic and geometric background are referred to Green and Carroll (1976), or to part one of Van der Geer (1971) for a briefer and more elementary approach.

Finally, the observant reader may notice that the user is referred to throughout the book as 'she'. This is deliberate—it avoids the ugly 'he or she' form, and any remaining unease exemplifies just the point which professional guidelines on gender usage aim to make.

Tony Coxon
Cardiff, July 1981

Acknowledgements

My thanks are chiefly due to Clyde Coombs and James Lingoes of the University of Michigan; to Doug Carroll, Joe Kruskal and Roger Shepard of Bell Laboratories (now at Stanford University); to Eddie E. Roskam of Nijmegen University and to Forrest W. Young of the University of North Carolina. Thanks are due in particular ways to Phipps Arabie, Scott Boorman, Ingwer Borg, Mike Burton, Frank Critchley, Mark L. Davison, John C. Gower, Paul E. Green, Joe H. Levine, Ed Schneider and Ian Spence, who at sundry times and in divers manners have affected my thinking on MDS.

To my collaborators, co-investigators and colleagues in the MDS(X) project my especial thanks are due: to Charles Jones now of McMaster University, with whom I have worked and taught in MDS for many years, as also to David Muxworthy and Mike Prentice of Edinburgh and Steve Tagg of Strathclyde University. To Peter Davies I am specially grateful—the extent of his assistance, advice and contributions to this book are incalculable. Thanks are also due to Roy Omond, our original programmer, to Peter Sykes, our current one, and to Monroe Wright and Erik Skovfoged who have worked at various times on the project. I also want to give particular thanks to Myrtle Robins our long-suffering and efficient project secretary and decoder of hieroglyphics who has typed and mothered the manuscript through its various drafts to its final form; to Margaret Simpson who has in various secretarial and organisational ways assisted us, and to Joan Ryan who typed the bibliography.

Several colleagues and friends have read successive drafts of this monograph and have commented liberally on them. I am especially grateful to Monnie Williams and Wijbrandt van Schuur who have kept up a regular and invaluable succession of helpful, detailed (and largely heeded) comments, criticisms and suggestions. Thanks are also due to J. van Deth, R. Ecob, J. James and C. Richardson who have read particular drafts, and to the very many students whom I have taught in MDS courses, especially at the European and International Consortia on Political Research, and at seminars at the Zentralarchiv für Empirische Sozialforschung in Cologne and at the University of Groningen. I should also like to thank those who have organised such courses, arranged the distribution of the MDS(X) package and provided the back-up without which the project and these texts would not have been possible: Maria Wieken and Gerhard Held of the Cologne Zentralarchiv and Marjorie Barritt deserve my special thanks in this regard. As our agent, Caroline Dawnay has provided magnificent help, advice and support in the production of these volumes for which we are very grateful, and thanks are also due to David Hill and Robin Frampton who have seen the books in to production. Our particular thanks go to Anne Murcott for her assiduousness, care and eye for detail skills have been invaluable in the extensive help she has given in proof-reading and indexing. David Parry has also made many valuable contributions especially in the final stages.

The author and the publishers would like to thank the following authors and publishers for their kind permission to reproduce their material in this book (numbers of the pages on which this material occurs are given in brackets): J. Alt, B. Sarlvick, I. Crewe and Elsevier Publishing Company (198); J. D. Carroll and the Center for Advanced Study in Behavioral Sciences, California (227); J. C. Gower and *Psychometrika* (206); S. Levy and Elsevier Scientific Publishing Company (99); S. Rosenberg, A. Sedlak and Academic Press, New York (96, 107–108); E. E. Roskam (55); Gilbert Shelton and Knockabout Comics, London (xiv); R. N. Shepard and Bell Laboratories, New Jersey (6, 7); R. N. Shepard and *Journal of Mathematical Psychology* (59, 67); R. N. Shepard and *Science* (250, 253); I. Spence, J. Graef and the Institute of Multivariate Behavioral Research, Texas (85).

List of Figures

List of Tables

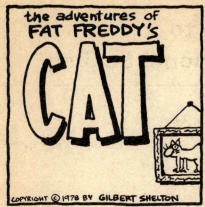

Introduction to Multidimensional Scaling

Ye konne by argumentes make a place
A myle brood of twenty foot of space

CHAUCER (The Reeve's Tale)

Multidimensional scaling (hereinafter abbreviated to MDS) refers to a family of models by means of which information contained in a set of data is represented by a set of points in a space. These points are arranged in such a way that geometrical relationships such as distance between the points reflect the empirical relationships in the data. For example, the complex associations between a set of variables which is contained in a matrix of correlations can be represented spatially by portraying each variable as a point, placing them in such a way that the distances between them reproduce the numerical value of the correlation coefficients. Thus, a picture of the data is produced which is much easier to assimilate (visually) than a large matrix of numbers. It may well also bring out features of the data which were obscured in the original matrix of coefficients.

There are three crucial things which the researcher must consider in using multidimensional scaling:

(i)   *the data* which provide the information which is to be represented;

(ii)   *the model* which interprets the data in a particular way, e.g. as giving information about the relative proximity of the objects;

(iii)   *the transformation* which specifies which information in the data is to be preserved in the solution—e.g. in the basic non-metric MDS model we seek to reproduce only the rank-order of the entries in the data matrix.

These three characteristics will be used time and again to distinguish the varieties of MDS.

1.1   An Example

Suppose a sample of subjects in a survey has been asked to compare a set of eight legal offences, and to say for each one *how unalike (or dissimilar) it is in terms of its seriousness* compared to each of the others in turn. The resulting data are presented in Table 1.1: each entry in the table tells us the percentage of respondents who judge the offence read across the row as being 'very dissimilar' to the offence read down the column, in terms of their seriousness (thus, 63.4 per cent say that receiving stolen goods and perjury are viewed as being very unalike in terms of their seriousness). It is difficult to take in all the information contained in the 28 coefficients, though some features are fairly obvious—libel and perjury are clearly

	Offence	1	2	3	4	5	6	7	8
1	Assault and battery	(0)							
2	Rape	21.1	(0)						
3	Embezzlement	71.2	54.1	(0)					
4	Perjury	36.4	36.4	36.4	(0)				
5	Libel	52.1	54.1	52.1	0.7	(0)			
6	Burglary	89.9	75.2	36.4	54.1	53.0	(0)		
7	Prostitution	53.0	73.0	75.2	52.1	36.4	88.3	(0)	
8	Receiving stolen goods	90.1	93.2	71.2	63.4	52.1	36.4	73.0	(0)

Table 1.1   *Percentage of subjects saying offences are very unalike in their seriousness*

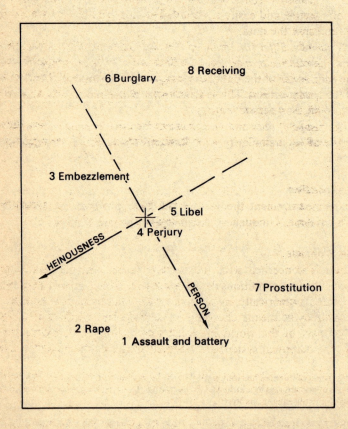

Figure 1.1   *Seriousness of offences: 2-D scaling of data of Table 1.1*

thought to be very much alike in the degree of their seriousness (as indicated by the low degree of dissimilarity), whilst almost every respondent agrees that receiving stolen goods and rape differ considerably.

The information can be represented—and absorbed—much more easily if we now proceed to an MDS analysis. The data are first treated as distances, and then scaled geometrically in two dimensions, as is done in Figure 1.1. In this diagram (termed a configuration, or pattern, of points) the more dissimilar a pair of offences is judged to be, the greater the distance between the points representing them. Thus, as we have noted above, the distance between points 4 and 5 is the smallest, and that between points 1 and 8 is the largest. Procedures for interpreting such configurations are introduced in Chapter 4, but without prejudicing that account, we may note that two fairly independent contrasts seem to underlie the configuration—what might be called the relative heinousness of the offence (rape, embezzlement *vs* receiving, soliciting), and the extent to which a crime is against the person rather than against property (assault, soliciting *vs* burglary, embezzlement).

This example, although based on fictitious data*, allows a number of points to be noted:

(i) *MDS is primarily concerned with representation*—in this case with the production of a simple and easily assimilated geometrical picture where distances are used to represent the data.

(ii) *MDS models differ in terms of the assumptions they make about how important the quantitative properties of the data are.* In the example above, it is in fact only the *rank order* of the data percentages which is matched perfectly by the distances of the configuration. This is an example of ordinal scaling or, as it is more commonly termed, non-metric scaling.

(iii) *A wide range of data and measures can be used as input.* Any data which can be interpreted as a similarity or a dissimilarity measure are appropriate for scaling analysis.

1.2 Representation

Contemporary measurement theory sees the basic problems of measurement as being: Representation, Uniqueness, Meaningfulness and Scaling.†

1.2.1 Measurement

Representation is concerned with the logic of measurement, that is, with the necessary and sufficient conditions data must satisfy if the property they measure is to be represented numerically, or in some other mathematical system. In this account, measurement means choosing an appropriate mathematical model to represent that part of the world in which one is interested, ensuring that the properties of the empirical system are reflected in the mathematical system.

---

*Adjudged seriousness of offences has long been studied by scaling methods (Thurstone 1927; Coombs 1967), and the use of percentages as a measure of association also has a long history. This artificial set of data is used again in subsequent chapters.

†See Suppes and Zinnes 1963; Adams 1966; Ellis 1966; Krantz et al. 1971, for definitive accounts of modern measurement theory, and Coombs et al. 1970, part 1, for a lucid and simplified account.

In the context of MDS, the range of models available for representing a particular set of data is often wide. Any particular model or scale which we choose for representing our data consists, in essence, of a theory about the empirical system we are studying, and like any other theory it necessarily embodies restrictive assumptions. We buy information about our data by making such assumptions, but at a cost: more complex and quantitative measurement demands more restrictive assumptions.

In point of fact, the assumptions which measurement models make are rarely met in any real set of data, and the strict representationalist position serves more as a future hope than as a present help.

Index measurement

Strictly speaking, the failure of a set of data to satisfy the axioms of a measurement model means that, at best, one has achieved 'index measurement' (cf. Dawes 1972, p. 91 et seq., pp. 146–8), i.e. the properties of the numbers assigned in measurement do not necessarily imply anything about the properties of the objects being indexed, unlike in the paradigm of representational measurement where there is a direct correspondence between the properties of the empirical system and the numerical measurement system.

The logical force of the representational account is such that a single error would falsify the model. Faced by such a strict requirement one needs either to add a theory of error to a measurement model (indicating how many exceptions can be accepted before the model is abandoned), or view representational measurement simply as an ultimately desirable goal. Moreover, given the heavy dependence of empirical research on index measurement—and on the 'ubiquitous rating scale' in particular*—it is hard to maintain so uncompromising a position. For some models in MDS, representational measurement is possible and tests can be made on the data. For others this is not so, but *so long as unjustified measurement claims are not made for the solution*, and so long as tests of axioms are made where they exist, it is entirely reasonable to use procedures such as MDS. In the case of the distance model, there is further pragmatic justification. When MDS is used as a method of data reduction and/or a graphical device for displaying structure in data (see Everitt 1978), its use is very like any other form of descriptive statistics used to portray similarity, such as histograms and scatter diagrams, where we use our intuitions and knowledge about our three-dimensional world to grasp information about the data we analyse. Whilst this may not amount to representational measurement or strict inference, it rightly counts as exploratory analysis, or what Gnanadesikan (1973) refers to as 'informal inference'.

1.3 Uniqueness and Scale Type

Once a property or relation has been measured, there is still a degree of arbitrariness in the actual scale values assigned. Any *reallocation* of numbers which preserves the crucial information in the original scale values is termed a transformation, or a re-scaling. The so-called uniqueness problem can be reduced to a simple question: what sort of transformation is legitimate, in the sense that it

---

*Dawes (1972, p. 96) cites the fact that 61 per cent of 172 experimental studies in a year's issue of the *Journal of Personality and Social Psychology* used rating scales as the dependent variable, and in over a third of these articles, rating scales were the *only* data used for the dependent variable.

leaves the properties of the original relation (or property) unchanged? Put slightly differently: how unique are the original scale values?

Based upon the notion of 'legitimate transformation', a number of typologies of the 'degree of uniqueness', or 'levels of measurement' have been proposed. The most long-lived classification is that of Stevens (1946) which distinguishes nominal, ordinal, interval and ratio levels of measurement. This simple typology has certain appealing features—it forms a cumulative hierarchy, in the sense that each level includes as a subset those beneath it, and it includes those levels which are of most concern in MDS, namely ordinal (which is the usual level of input data) and ratio (which is the level at which distance is measured).

The original typology has been subject to a good deal of methodological criticism and elaboration, especially in the form of adding further 'non-metric' scale types between the nominal and interval levels and in distinguishing between different types of ordinal scale. For our purposes it is useful, first, to extend the basic typology to cover all those scales mentioned in this book and secondly, to make some important distinctions between types of ordinal scale.

Shepard (1972, p. 7) has provided a very useful extension of this basic typology which is reproduced with slight changes in Table 1.2 and Figure 1.2. Our interest will focus principally upon the variants of the ordinal scale, and to this we now turn.

1.3.1 *Types of ordinal scale*
It is important to distinguish strict, partial and weak order.

(i) *Strict ordering (alternatively called a strong order or chain; Stevens' ordinal scale)*
Information is complete (i.e. exists about *all* pairs of objects). The pairwise data are irreflexive, asymmetric and transitive and thus can be represented as a single ordering.

(ii) *Partial order*
Information is incomplete and in such a case the objects cannot be reduced to an unambiguous single ordering.

Partial ordering often results from missing data, and since this is a fairly common occurrence in MDS analysis we shall need to know about such things as how many missing data can be tolerated before a solution becomes unstable, and whether systematic missing data (for example, most of the information referring to one point) is more dangerous than more randomly occurring omissions. These points are discussed below.

(iii) *Weak order*
In this case, information is complete, but some objects occur at the same level of ordering instead of being each at a distinct level, as in the strict order.

Weak orders often arise in preference judgments, where, because two objects are so similar, a person may say both that she prefers *a* to *b* and later that she prefers *b* to *a*. Usually this would be considered as an 'indifference' relation, meaning that the difference between the two objects is too small to be significant or reliable (see Luce's (1956) theory of preference semi-orders for such a model).

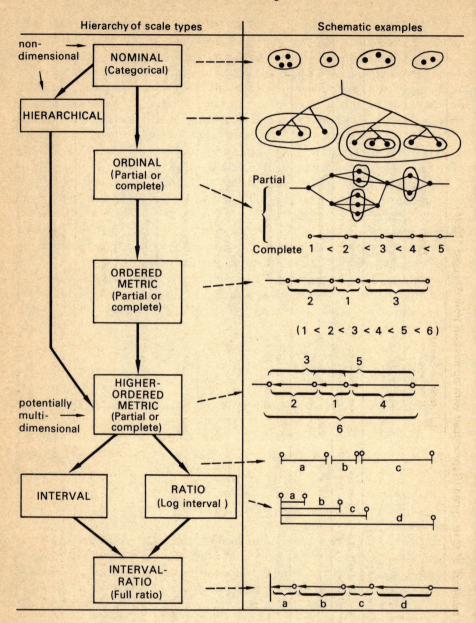

Figure 1.2   *Diagrammatic classification of scale types*

More commonly, weak order arises because only a few values are used in a rating scale. The significant questions are: how many *distinct* values must appear in a matrix for a solution to be possible; and, are tied values in the data to be re-scaled into tied values in the solution, or not? (In MDS, this is referred to as the secondary and primary approach to ties). These questions are taken up in section 3.2.2.

Table 1.2 Shepard's typology of scale-types (Reproduced with permission from Shepard 1972a)

Type of Scale	Given Information (for p objects)	Added Information (over preceding scale type)	Permissible Transformations of Scale	Examples
NOMINAL (Categorical)	Assignment of each of the p objects to a category in an unordered set of mutually exclusive, labelled categories	Assignment to labelled categories	Permutations (of the category labels)	Numbering of football players
HIERARCHICAL*	Assignment of each of the p objects to a terminal node of a graph-theoretic tree	Grouping of categories into super-categories	Interchanges of topologically equivalent branches of tree	Biological taxonomies
ORDINAL	Rank ordering (complete or partial) of all p objects	Rank order of (nominal) categories	Monotone transformations: $x' = f_{monotone}(x)$	Mohrs' scale of hardness of minerals
ORDERED METRIC	Rank ordering (complete or partial) of the distances between the two objects in all $(p - 1)$ adjacent pairs	Rank order of distances between adjacent points	Overall linear transformation, plus considerable local perturbations	Coombs' (1950) 'scaling without a unit of measurement'
HIGHER-ORDERED METRIC**	Rank ordering (complete or partial) of the distances between the two objects in all $p \times (p - 1)/2$ pairs	Completion of rank order for all pairs of points	Overall linear transformation plus limited local perturbations	Complete proximity matrix
INTERVAL	Specification of the numerical values of the distances between the objects in all $(p - 1)$ adjacent pairs	Numerical values of intervals	Linear transformations: $x' = a + bx$	Fahrenheit, or Celsius scale of temperature
RATIO (Log-Interval)	Specification of the numerical values of the ratios between the objects in all $(p - 1)$ adjacent pairs	Numerical values of ratios	Power transformations: $x' = ax^b$	Scales based on direct magnitude estimation
INTERVAL-RATIO (Full Ratio)	Specification of the numerical values of distances of all n objects from a unique zero point	Zero point	Values intervals	Similarity transformations: $x' = ax$	Absolute scale of temperature

*The hierarchical scale is basically nondimensional, though it may be consistent with a representation of the same points in a one- or higher-dimensional space.

**The higher-ordered metric scale is unique among these in that it is potentially multidimensional. We can arrange four points. A, B, C, D, on a one-dimensional straight line in such a way that the magnitudes of the six interpoint distances have the rank order $AB < CD < BC < AC < BD < AD$. Thus:

A ——— B —— C ———— D

But if we merely reverse the order of the two distances AC and BD, we obtain the order $AB < CD < BC < BD < AC < AD$ which can be accommodated only in a two-dimensional space.

(iv) *Ordered metric scales*

A further addition to the original Stevens' typology is the 'ordered metric' scale between the ordinal and interval level of measurement (Coombs 1950) for the situation where there is both an ordering between the points, and also information on the order of the magnitude of at least some of the constituent intervals. This scale type is termed 'ordered metric' because such pairwise differences provide information on the *order* of the magnitude of the differences.

The distinctions between these various types of ordering are important in understanding a number of points which arise in non-metric MDS. In particular they become crucial when considering what aspects of order information in the data the user wishes to see represented in the solution (see 3.2.3).

1.4 Ordinal Scaling

In the basic MDS model which we shall be discussing, the scaling problem consists of locating a set of points (one to represent each object) in a space in such a way that the information in the data is represented as faithfully as possible by the separation or distance between the points. Information of a purely ordinal kind—such that rape is considered more serious than larceny—places very few restrictions on where the points can be positioned in conformity with the data. By contrast, a matrix of dissimilarity coefficients can be interpreted, as we shall see, as providing ordered metric information about the rank order of the *differences* between pairs of objects—for instance, that burglary and rape differ more in seriousness than perjury and libel. In this case the distances of the solution will be required to conform to the same rank order as the entries in the data matrix. This imposes more severe restrictions on the positions which the points can occupy in the solution space. The point locations thus become more and more fixed as more order restrictions are applied, and this results in a more stable configuration.

In this scaling process, which is described in considerable detail in section 3.5, two important considerations arise which are concerned with problems of levels of measurement. First, the basic input data are usually in the form of a symmetric matrix of similarities or dissimilarities measures. If the researcher wishes only the *order* of the entries in the matrix to be used in arriving at the solution, then this is called non-metric (Shepard 1962, Kruskal 1964) or ordinal scaling (Sibson 1972).

A second way of looking at this scaling process is to take the original data as initial values, and to try to find a permissible transformation into a 'better behaved' set of re-scaled values, such as Euclidean distances. In the present case, the transformation which is considered legitimate is ordinal or monotonic, since only the order of the data is being preserved. It is a little harder to define precisely what 'better behaved' means. Loosely, it means that the re-scaled values will have more desirable (and preferably metric) properties than the input data. In particular, if the original data can be ordinally re-scaled in such a way that they become Euclidean distances, then it will be possible to represent the data as a configuration of points in a Euclidean space.

There is a danger that such a procedure may appear to be a magical device for turning ordered metric data into ratio level data (distances). In case the researcher suspects that MDS consists of a statistical sleight of hand, it should be emphasised that a *perfect* ordinal transformation will rarely be found in practice. Instead, the researcher will have to decide how great a departure from a perfect ordinal rescaling can be tolerated before deciding that the data cannot be legitimately transformed (scaled). In Coombs' terms, the lack of fit between the original data and the re-scaled values is one of the costs that has to be paid to obtain the solution. For non-metric scaling, the cost is a good deal less than it is for metric scaling, where the data are assumed to be at least interval level.

1.5 Applications of MDS

Articles and books using MDS are now appearing at an ever-increasing rate and in a growing number of disciplines. By and large, the area of application has its historical origin in psychology, and has diffused principally into the social sciences.

Even ten years ago it was possible to publish an extensive—and probably a fairly exhaustive—bibliography of over 300 MDS applications in the Bell Bibliography (1973). It is now becoming a Herculean task even to keep track of applications within a single discipline.

Probably the best and most accessible source of new publications is contained in the annual literature search produced by the Classification Society, which uses a 'key' of classic MDS articles to identify articles which cite one or more of them.

In our book, a number of relevant applications will be included in each chapter and further examples are included in the *User Manual* reports for each program in the MDS(X) series.

Caveat lector

Since MDS can be used on a very wide range of data, Chapter 2 is devoted to discussing the general notion of a 'distance measure' and providing instances of the wide variety of types of data which may be used as input to MDS. *The user will find it useful to skip Chapter 2 at first reading and come back to it at a later stage.* Initially it may well seem to be a long catalogue of possible measures. However, Chapter 2 should be read at some point, since it provides important clues about how virtually any sort of data can be viewed as giving information suitable for MDS analysis.

2 Data: Measures of Similarity and Dissimilarity

*You shall have true and correct weights and true and correct measures, so that
you may live long in the land which the Lord your God is giving you*
DEUTERONOMY **25**, 15 (New English Bible)

A very wide range of types of data and of measures of association can be
interpreted as providing information about how similar or dissimilar objects are to
each other. In turn, the idea of similarity between two objects lends itself readily to
interpretation in terms of proximity of two points. Conversely, the more dissimilar
the objects are, the more distant the corresponding points should be. The definition
of distance measures and the relationship between distances and scalar products is
presented in Appendix A2.1. *Readers unfamiliar with these notions are strongly
recommended to read this material before proceeding.*

Empirical measures
In MDS, a number of different names have been given to empirical measures which
are thought of as being estimates of distance (or its antonym, proximity). In this
text the generic term 'dissimilarity' will be used to cover any distance measure, and
'similarity' will be used to refer to a proximity measure. The empirical dissimilarity
measure between two objects j and k will be denoted: δ_{jk}, and any similarity s_{jk} can
readily be converted into a dissimilarity measure by reversing the values.

A large number of (dis)similarity measures can be used in MDS. There are two
basic types:

(i)   (Dis)similarity data obtained directly in the form of numerical or order
estimates ('direct data'). Much attitudinal and survey data are of this form.

(ii)   (Dis)similarity data based on aggregating direct data ('aggregate' or
'derived' measures). Usually, association and correlation measures are of this form.

Each type will be considered in turn.

2.1   Direct Data
The individual data collection methods have been well discussed by Coombs (1964,
ch. 2) in terms of the amount of information which can be obtained by a given
method (its channel capacity) and the amount of implied or repeated information
(redundancy) it contains, which can be used to check the subject's consistency. His
summary is well worth repeating:

> On *a priori* grounds we would expect that the higher the channel capacity the
> better, but this is certainly not true. A price is paid for data, not only in financial

terms but in wear and tear on the organism at source. A method with too high a channel capacity may, through boredom and fatigue, result in a decrease in information transmitted, through stereotype of behavior. Furthermore, the potential variety of messages from the organism may not be great, in which case a more powerful method is inefficient ... Ideally a method should be selected which matches the information content in the source but is not such a burden as to generate noise.

(ibid, p. 51)

To a large extent fatigue, stereotyping and inefficiency can be cut down by the use of experimental designs, which provide methods for selecting incomplete sets of judgments, preferably with desirable statistical properties. In Table 2.1 a brief summary of the more commonly used methods is presented, and references which contain information on incomplete designs are asterisked.*

2.1.1   Pair comparisons data

Collecting *similarity ratings of pairs* of stimuli from a set of subjects is a long-established and popular pastime, especially among social scientists, but there is no guarantee that subjects' use of rating categories bears any resemblance to the arithmetic properties which are often ascribed to them. In any event, researchers should take care to ensure that a sufficiently large number of categories is provided (Green and Rao (1970) provide further empirical support for Miller's (1956) 'magical number 7 ± 2' as an optimum number) and that subjects use a sufficiently wide range of them. Pairwise ratings are especially prone to 'response sets' or 'response styles' (Cronbach 1946, Rorer 1965) involving highly skewed distributions of ratings (Coxon and Jones 1978a, pp. 68–71, and 1979, T3.3 and T3.4), and wide variation in the number of categories which a subject uses.

Dominance judgments of pairs of stimuli are also fairly common. In this case the subject is presented with a pair of objects and asked to indicate which of the two possesses more of a given attribute (is heavier, louder, more prestigious, is preferred, is more sexy etc). The binary data generated in this way can be used to test the transitivity of each single subject's choice (see Kendall 1962, p. 144 et seq.), and are frequently turned into a rank-ordering if the choice is sufficiently consistent.

2.1.2   Partitions (sorting) data and hierarchies data

Any method by which a set of objects is divided into mutually exclusive and exhaustive categories constitutes a partitioning (or 'nominal scale'). The most commonly encountered forms of data collection which produce a partition are:

(i)   *Dichotomisation*   A set of objects is divided into *two* contrasting groups usually in terms of whether or not the objects concerned does, or does not, possess some specified property.

(ii)   *Trichotomisation*   Usually, the trichotomy consists of those objects which possess the property, those which do not, and those to which the property does not apply.

---

*Discussion of practical and methodological issues involved in these and other methods of data collection is contained in Shepard 1972c and in Coxon and Jones 1979, T2.4 and U1.4.

Table 2.1 Some direct methods of collecting dissimilarities data

Method	Presentation	Question Instructions	Implied Information on Distances	References
1 PAIR COMPARISONS	All pairs (j, k)	How dissimilar are j and k? e.g. on a 7-point rating scale from totally similar (1) to totally dissimilar (7)	δ_{jk} gives direct estimate of d_{jk}	*David 1963 *Spence and Domoney 1974 Green and Rao 1970
2 PARTITIONS (Sorting)	Subject divides stimuli into mutually exclusive and exhaustive categories or groups	Sort/divide stimuli into: (a) fixed number of categories (fixed sorting) (b) as many or as few categories as you wish (free sorting)	For any three stimuli, (j, k, l) $d(j, k) < d(j, l)$ if j and k are in the same category, and l is in a different category	Miller 1969 Anglin 1970 Jones and Ashmore 1973 Coxon and Jones 1978b, 1979
3 HIERARCHIES	Subject constructs hierarchical clustering of stimuli	(i) First choose the most similar (MS) pair, then (ii) Either add new stimulus, or begin new similar pair, then (iii) Either start new pair, or add stimulus, or merge existing cluster (see Coxon and Jones 1978b)	δ_{jk} is given by the level of hierarchy at which j and k are joined	Johnson 1967 Coxon and Jones 1978b, 1979 Fillenbaum and Rapoport 1971
4 RANKING	Subject i places stimuli $j, k, l \ldots$ m in rank order in terms of given criterion	What is your order of preference/similarity? Say: m, l, k, j	$d_{im} < d_{il} < d_{ik} < d_{ij}$	Coombs 1964 *Durbin 1951
5 TRIADS	All triads (j, k, l) of stimuli	(a) Which is the most similar pair? Say: (k, l) (b) Which is the most similar (MS) pair, and the least similar (LS) pair? Say: MS is (j, k) and LS is (k, l)	$d_{kl} < d_{jl}$ and $d_{kl} < d_{jk}$ $d_{jk} < d_{jl} < d_{kl}$	*Burton and Nerlove 1976 *Cochran and Cox 1951 Torgerson 1958
6 TETRADS	All pairs of pairs $((j, k)$ and $(l, m))$ of stimuli	Which pair is the more similar? Say: (l, m) MS than (j, k)	$d_{lm} < d_{jk}$	*Cochran and Cox 1951 Torgerson 1958

*References marked with asterisk contain information on incomplete designs to reduce number of presentations.

(iii) *Fixed sorting*   The objects are allocated to a pre-specified number of categories.

(iv) *Free sorting* (or 'own categories')   The objects are allocated to a set of categories, but the number of categories is not specified, and can in principle range between one in which all the objects are lumped together in one category and the case where each object forms its own category (Arabie and Boorman 1973).

Such data occur in a wide variety of disciplines, and are especially prevalent in cognitive studies, such as folk taxonomies in anthropology (Tyler 1969), psycho-semantics (Miller 1969), subjective classification in sociology (Burton 1972, Coxon and Jones 1978b), and in personal construct theory in psychology (Bannister and Mair 1969). Some examples of free sorting from a number of different disciplines are:

the co-existence of plant species (objects) within chosen field sites (categories);
the sleeping habitat (e.g. trees, which here constitute the 'categories') of a group of monkeys (the 'objects');
the co-location of a set of artefacts (objects) within a set of neolithic graves (categories)
the sorting of a set of words (objects) into piles (categories) in terms of their similarity of meaning;
the co-occurrence of themes (objects) within a set of documents or sentences (categories).

Partitions (nominal) data are usually pre-processed before being scaled. Often, each partition is turned into a matrix of co-occurrence between the objects, where an entry of '1' in δ_{jk} means that objects j and k both occur in the same category, and '0' otherwise, and the individual matrices are then summed. The analysis of aggregate co-occurrence matrices will be discussed in 2.2.3.3. Another alternative, when interest focuses chiefly on how similar partitions are to each other, is to compare them two at a time, and a dissimilarity measure is then computed for each pair. This is discussed under 2.2.3.5.

Hierarchies data, like partitions produced from the method of sorting and the 'tree-construction' method employed by Fillenbaum and Rapoport (1971, pp. 10–11, 15 et seq.), can be thought of as another way of getting a full set of similarity judgments from a subject without making the task too strenuous. The analysis is similar to that used for comparing partitions (see Coxon and Jones 1978b, ch. 7).

2.1.3 *Rankings and ratings data*

Rankings correspond at the individual level to the ordinal scales of measurement. They are one of the most popular methods of data collection in the social sciences, yielding a considerable amount of information at relatively little cost. The basic form (strict order ranking) has usually been obtained by presenting the subject with a set of objects or stimuli and defining a criterion by which the subject is to make his or her judgments. The subject is then instructed to choose the object which is highest on the specified criterion, followed by the next highest object and so on until a complete order is obtained, with no ties allowed. Several variants exist, especially the *weak order* (where objects may be tied, or treated as equal in terms of the criterion), and the *partial order* (where the subject is allowed to omit some objects). As we have seen, rankings may also be derived from a set of pair comparison dominance judgments if the subject is sufficiently consistent.

Ratings are often obtained by asking the subject to assign a number to each object in such a way that the magnitude of the number reflects her estimate of the extent to which it possesses the attribute defined by the criterion. Common variants of this are where the subject is asked to 'mark each stimulus out of 100' (percentage or 'thermometer' rating, so named because the rating scale is represented to the subject in the form of a thermometer scale), and the 'graphic scale', where the subject positions (say) an arrow to mark her rating of the object, and the investigator then measures the location with a ruler.

The assumption is that each rating judgment is made in relation to the other stimuli which are presented, but in accordance with some absolute standard. This method is known in the psychometric literature as magnitude estimation or direct estimation (Stevens 1966). Quite often, the researcher decides to ignore the numerical information and turn the subject's ratings into a rank order.

Both rankings and ratings give rise to 'rectangular' or 'conditional similarity' data—that is, each subject's rankings or ratings are not considered to be comparable directly to those of other subjects.

2.1.4 Triadic data

Triadic data (where the subject selects the most similar pair out of three objects) are very commonly collected by psychologists and others using Kelly's 'repertory grid analysis', which is also used to elicit the constructs which people use in interpreting their social world (Kelly 1955; Bannister and Mair 1969). A common, but dangerous practice is to turn triadic data of this sort into 'vote frequencies', counting the number of times a particular pair is judged more similar than another. Roskam (1970) has shown that this practice can badly distort the information in the data, and should be avoided. Another variant of triadic data occurs where the subject is presented with three objects and then asked to select the most similar pair *and* the least similar pair. In this variant, much more information is obtained than in the first case. Suppose the subject is given the triad (A, B, C) and chooses AC as the most similar and AB as the least similar pair. This implies that $d(A, C) < d(B, C) < d(A, B)$. If the subject is only asked to select the most similar pair we should only be able to infer in this example that $d(A, C) < d(A, B)$ and $d(A, C) < d(B, C)$ but we could infer nothing about the relationship between (A, B) and (B, C) (see 6.1.3).

Although triadic data collection appears to be a powerful and efficient method, and one which is well suited to obtaining personal constructs, it is rarely so in practice. Even when incomplete designs are used for reducing the number of judgments, subjects tend to find the task tedious, and their constructs often become highly stereotyped. Moreover, when several subjects' data are pooled, they cannot usually be represented well by scaling models (see Coxon and Jones 1978a, p. 66 and 1979, T3.1 and T3.11).

2.1.5 General issues in direct data collection: numbers and complexity

By and large, the most frequently encountered problems in collecting and scaling individual sets of data centre around the number of objects and the cognitive complexity of the data collection task. Usually a trading relationship exists between these two characteristics: the more complex the task, the fewer the number of objects that can be employed. Statistical designs for reducing the number of objects to be presented to the subject are certainly useful and are often an elegant solution. But balanced incomplete designs only exist for some types of data collection and for certain numbers of objects, and the user must be prepared to use more practical and rough-and-ready procedures if the research problem really merits it. This topic is discussed further in sections 7.2.1.1 and 7.5.5.3.

The cognitive complexity of tasks poses rather different problems. Experience shows that one's presuppositions about the ease and speed with which a data collection task is completed can be misleading. For instance, ranking turns out to be a far more difficult task than rating, since constant re-ordering and comparison is necessary to yield an ordering, whereas a rating can be made easily, and each object can be judged separately and without repeated comparison.

In a similar way, triadic judgment seems very well suited to eliciting constructs (or bases of judgment) *and* to producing fairly complex (ordered metric) data in a simple format. And so it is, methodologically speaking—except that subjects often find it an extremely wearisome task, and tend to 'lock in' on a single construct (Coxon and Jones 1979, T2.4). By contrast, hierarchy construction—a very complex and time-consuming task—was found to be an interesting and rewarding task yielding rich and reliable data.

Rao and Katz (1971) provide more systematic evidence of such factors in a study of the effectiveness of seven data collection methods, evaluating each method by a simulation method using data from a known configuration. They find that ordering and selection (or 'pick k out of n': Coombs 1964, ch. 2) methods such as pair comparisons, triads and tetrads produce better recovery of the original configuration than sorting methods, but that hierarchy construction is superior to the other sorting procedures.

2.2 Aggregate Data

Most frequently, the measure of dissimilarity used as input to MDS programs is an aggregate measure (summed over individuals, replications, times, locations etc.), and usually it is also an index of association (typically a measure of correlation or contingency).

As a methodological principle, the inspection of individual differences should always precede aggregation. If the individuals (or other units of analysis) differ systematically among themselves with respect to variables of interest, then such information is lost upon aggregation. Indeed, if data are aggregated, one can never know whether or not such differences even exist. Moreover, if significantly different subgroups do exist then any averaged information will be an artefact and will not reproduce the characteristics of either group accurately. There is always the danger of 'piecemeal distortion' as well. If the data referring to a given unit or individual is complex, then 'local structure' (interrelationships within parts of the data) can be lost entirely when the components are aggregated. It is sometimes argued that individual variation is simply unnecessary noise (or error) which will cancel out on aggregation. Perhaps it will, but such a belief requires a degree of well-behavedness on the part of error that, however commonly assumed in social science modelling, is scarcely realistic and should at least be investigated.

In any event, a cautious approach is to be preferred. One way is to concentrate attention initially on examining each subject's data structure (meaning here simply 'set of pairwise judgments', or 'rankings', or 'triads', or whatever), and then compare the entire structure of different individuals before examining the aggregate structure of the objects.

In order to simplify matters, we shall assume that the basic data matrix, X, from which these summary measures will be computed is a rectangular matrix, with N rows (usually representing individual subjects or units) and p columns (each representing a separate variable or attribute) and whose element x_{ij} gives the value on variable j for individual i.

Level of measurement

Since each measure of association takes into account the level of measurement of each of the variables involved, it is convenient to distinguish measures intended for interval, ordinal and nominal data (counting dichotomous and co-occurrence measures as special cases of nominal data). The meaning of the word 'association' changes according to the level of measurement, and so it makes sense to compare directly only those measures which summarise data at the same level. Attention will be restricted to *symmetric* measures of association, although asymmetric measures can be represented in scaling models (see below).

R and Q analysis

Measures of association also differ as to whether they summarise the similarity between pairs of variables (columns of the data matrix), or between pairs of subjects (rows of the data matrix). The first type is often termed R-analysis, and the second Q-analysis. In some cases a measure can be used in either way, but usually a particular measure is designed for one form of analysis rather than the other.

The list of measures presented in the subsequent sections makes no claim to be exhaustive; measures are chosen either because of their obvious suitability in representing similarity for scaling models, or because they are in common usage among behavioural scientists. More extended treatment of association measures is provided in Galtung (1967, pp. 205–33), Loether and McTavish (1974, pp. 185–262), Blalock (1972, chs. 13, 15 and 18) and Wishart (1978, chs. 28 and 29).

2.2.1   Interval level measures

The most commonly encountered measures of similarity for interval (and higher) level data are the product moment (PM) family of coefficients, each of which can be used in either R- or Q-analysis mode. Product moment measures are all basically *vector* measures (see Appendix A2.1) where the similarity between two variables (for R-analysis) or subjects (for Q-analysis) is represented by the combination of

(a)   the length of the vectors, and
(b)   the product (inner, or scalar product) of the two vectors, represented by the size of the angle separating them.

The most familiar and frequently used product moment measures of similarity are *covariance* (where the scale units of the variables enter into the assessment of similarity) and the Pearson product moment *correlation coefficient* (where the variables are standardised to unit length, and only the angular separation is considered). In neither case does the measure represent *distance* between the variables, although it is possible to convert vector separation (product moment) measures into distances (Appendix A2.1), and vice versa (Appendix A5.2). Vector and distance representations of similarity are described and compared in Appendix A2.1, where it is shown that since the correlation coefficient is a monotonic transformation of distance, it may be used directly in the basic *non-metric* MDS distance model.*

The basic matrix of scalar products (the product moment matrix) gives the information from which variance, standard deviations, covariances and correlations are produced, and is widely used in descriptive and multivariate statistics. From the data matrix **X**, two related product matrices can be formed:

the *minor* product matrix, **X'X**, a symmetric matrix of order p, whose entries give the scalar products (sum of squares and cross products) *between the variables* (R-analysis);

the *major* product moment matrix, **XX'**, a symmetric matrix of order N, or the scalar products between the subjects (Q-analysis or profile analysis).

Both matrices have a number of important and desirable statistical properties in common (see Green and Carroll 1976, p. 227 et seq.). In particular, they are of the same rank, which in the MDS context means that the data can be fully represented in a space of at most $m - 1$ dimensions, (where m is the smaller of N and p).

A number of variants of the basic PM matrix are in common use as measures of similarity:

1   Deviate PM matrix
Where the original data values have been 'centred', by having the column mean subtracted from each variable value, thus forming the matrix $\mathbf{X}_d$ of 'deviate scores': $x_d = (X_{ij} - \overline{X}_j)$. This has the effect of removing the overall average effect of each variable. (In the case of Q analysis, read 'subject' for 'variable', and 'row' for 'column' in the previous—and subsequent—sentences). The PM matrix formed from the deviate matrix is often termed the matrix of corrected (or deviate) squares and cross products (CSCP).

2   Dispersion (variance-covariance) matrix
When each entry in the deviate matrix is divided by N (or by p in Q-analysis), the PM matrix formed from it contains *variances* (averaged sum of squared *deviations*) in the diagonal elements, and *covariances* (averaged sum of corrected cross products) in the off-diagonal elements.

In both the deviate and dispersion PM measures, the units in which the original variables are scaled contribute directly to the overall measure of similarity (so that measurement of height in terms of metres will produce drastically reduced similarity values compared to measurement in centimetres).

---

*Vector separation and distance are not *linearly* related, so conversion between the two types of measure is necessary in metric scaling, and is sometimes an option provided within computer programs, such as INDSCAL.

3   Correlation matrix

If each variable is standardised (i.e. the score is centred and the resulting deviate value is then divided by the standard deviation of the variable) to normal scores $z_{ij} = (X_{ij} - \overline{X}_j)/\sigma_j$, the resulting PM matrix contains ones in the diagonal (representing the total variance), and the Pearson correlation coefficient r_{ij}, in the off-diagonal elements.

In calculating the correlation coefficient, each variable has been reduced to a common unit of measurement (its standard deviation) and differences in variance between the variables have hence been removed.

A useful comparison of these PM measures is provided by Skinner (1975), who analyses them in terms of three independent components:

1   *elevation* (m_j)—the average or mean effect of the variable j, so named because when a set of variable scores is drawn as a profile, the removal of the mean removes the elevation or average height of each variable;

2   *scatter* (s_j)—the dispersion, measured by the standard deviation of the variable j; and

3   *shape* (r_{ij})—measured by the correlation coefficient, representing simply the angular separation of two variables i and j.

In these terms, if original ('raw') data are converted into deviate data, information due to elevation is eliminated. Similarly, standardising the data has the effect of equalising the scatter of all the variables.

Each PM measure can then be broken down into its constituent components, which are related in the following way:

Product Moment Measure:	Components			Comment
1 Original (Raw scores)	$m_i m_j$ = Elevation	$+\ s_{ij}$ $+$ (scatter	$\times\ r_{ij}$ $\times$ shape)	Mixes all 3 components, merges scatter and shape
2 Variance-Covariance		$s_i s_j$ (scatter	$\times\ r_{ij}$ $\times$ shape)	Removes elevation, merges scatter and shape
3 Correlation			r_{ij}	Purely shape

Skinner's treatment is specially relevant in dealing with Q-analysis and profile data, when the researcher wishes to compare subjects in terms of their pattern of scores across a given number of items (test scores, semantic differential concept ratings etc.) Several of the programs in the MDS(X) series give the user the option of centring and standardising the subjects' data where this is appropriate.

2.2.2 Ordinal measures of association

Measures of ordinal (or monotonic) association address the question: To what extent do variables X and Y rank individuals (or whatever) in the same way? Perfect positive association occurs where individuals are ranked in the same order on both variables, and perfect negative association represents a total inversion in ordering. On this, all measures of ordinal association agree. But difficulties arise in giving meaning to intermediate degrees of association, and this is due to two factors:

(i) whether ranks are to be considered as numerical quantities (on which arithmetic operations may legitimately be performed) and

(ii) whether 'the same order' is to be interpreted in a strict or weak sense.

A number of relevant measures are summarised in Table 2.2.

Ironically, the earliest pioneering work on ordinal association (Spearman 1904, see Kendall 1962) produced a measure which is not strictly an ordinal measure at all!

Spearman's rho (Table 2.2(1))

This rank correlation coefficient is the product moment correlation between ranks, considered as integer quantities. The measure compares two rank orderings, and is based upon the (squared) *difference* in rank positions. It thereby measures not only the inversions which occur in two orderings but also the numerical size of the differences. Rho is invariant under linear transformations of the data but it is *not* invariant under monotone transformations, and it is therefore an interval level measure. It varies between -1 (when one ranking is the reverse of the other) and $+1$ (perfect agreement).

2.2.2.1 Ordinal measures based on inversions in rankings

Usually, ordinal variables are weak orderings consisting of a fairly small number of ordered categories, such as high, medium and low levels of motivation, or Likert's five response categories for attitude items (strongly agree, agree, not sure, disagree, strongly disagree). Consequently it usually happens that a large number of subjects share the same ordinal value, i.e. they are 'tied' with respect to the variable concerned. Different measures of ordinal association treat tied data in different ways.

The basic idea underlying genuine measures of ordinal relationships is that of comparing each pair of individuals on the two variables, and seeing how often they are ranked in the same way. An example will help clarify the concepts involved. Suppose we have data for ten individuals on the ordinal variables X and Y, (where H, M and L stand for High, Medium and Low respectively):

	Variables				Variables	
Individuals	X	Y		Individuals	X	Y
1	M	M		6	H	L
2	H	M		7	H	H
3	M	H		8	M	L
4	L	L		9	L	H
5	L	M		10	H	M

Name	Formula	Maximum	Zero	Minimum	Advantages	Disadvantages	Reduces to
1 SPEARMAN'S RHO (ρ)	$1 - \dfrac{6\Sigma d^2}{n^3 - n}$	+1	Both rankings identical	−1	Takes size of inversions into account	Treats ranks as interval quantities	—
2 GOODMAN AND KRUSKAL'S GAMMA (γ)	$\dfrac{C - D}{C + D}$	+1	Equal balance between concordant pairs	−1	Measures *weak* monotonicity (no reversals)		
3 KENDALL'S TAU (τ) (a)	$\dfrac{C - D}{C + D + T_x T_y + T_{xy}}$	+1	Equal balance between concordant pairs	−1	Best defined for *strict* rankings	Ties considered as errors	—
(b)	$\dfrac{C - D}{\sqrt{C + D + T_y}\,\sqrt{C + D + T_x}}$				Takes ties into account	Reaches maximum only for square tables	τ_a when no ties
(c)	Corrected version of tau$_a$: (Wilson 1974, p. 352) $\sigma_a\{n - 1)/n\}$ $\{m/(m - 1)\}$ where n is the number of cases and m is the number of cells in longest diagonal				Takes different number of ranks into account	'Quick and dirty way' of extending τ_b to non-square tables	σ_b when same number of categories
4 WILSON'S e	$\dfrac{C - D}{C + D + T_y + T_x}$	+1	Equal balance between concordant pairs	−1	Measures *strict* monotonicity		

Table 2.2 *Ordinal measures of association*

In comparing pairs of individuals there are five possibilities (see Wilson 1974 for an extended exposition):

1   *Concordant pairs* (C)   (*X* and *Y* order the individuals in the *same* way). If *i* is higher (lower) than *j* on *X*, then *i* is higher (lower) than *j* on *Y* (for instance, as occurs in comparing individuals 7 and 1, or 3 and 5).
2   *Discordant Pairs* (D)   (*X* and *Y* order the individuals in *opposite* ways). If *i* is higher (lower) than *j* on *X*, then *i* is lower (higher) than *j* on *Y* (e.g. as occurs in comparing 3 and 2, and 5 and 6).
3–5   *Tied Pairs*   (*X* and *Y* share at least one *tied* value)
    3   *Tied on* **X** (T_x)   (e.g. as occurs in 9 and 5, and 8 and 3)
    4   *Tied on* **Y** (T_y)   (e.g. as occurs in 8 and 4, and 2 and 5)
    5   *Tied on both* X *and* Y (T_{xy})   (identical values) (as occurs in comparing 2 and 10).

Measures of ordinal association all have the same basic form:

$$\frac{\text{(numerator)}}{\text{(denominator)}} = \frac{C - D}{N}$$

where $(C - D)$ is the difference between the number of concordant and discordant pairs, and N is the number of pairs which are considered to be relevant to the measure. (The denominator changes from measure to measure).
In terms of the numerator, the measures are either

weakly monotonic, allowing ties to count as concordant pairs, so that if *i* is higher (lower) than *j* on *X*, then *i* is at least as high (low) as *j* on *Y*; or
strictly monotonic, insisting that both inequalities *and* ties must be matched, so that:
    if *i* is higher (lower) than *j* on *X*, then *i* must be higher (lower) than *j* on *Y*, and
    if *i* is tied to *j* on *X*, then *i* must be tied to *j* on *Y*.

Goodman and Kruskal's gamma (Table 2.2(2); Goodman and Kruskal 1954)
This widely-used index measures *weak monotonicity* between two variables, and was expressly designed for summarising cross-tabulations of data. It is defined as the ratio of the *difference* of concordant and discordant pairs to the *sum* of concordant and discordant pairs. It therefore completely ignores ties on both *X* and *Y* and is described by Wilson (1974, p. 331) as the 'measure of the extent to which the data fit a 'no reversals' (weak monotonicity) hypothesis'. In the case of 2 × 2 tables, gamma reduces to the Q-coefficient, discussed below under 'dichotomous measures'.

Kendall's tau measures (Table 2.2(3); Kendall 1962)
The tau measures have been described as 'coefficients of disarray', and are also based on the difference of the number of concordant and discordant pairs. They differ from gamma in that they expressly take into account *all* pairs, whether tied or not. They are therefore measures of the strong monotonicity hypothesis and can also be interpreted in terms of the number of interchanges necessary to transform one ranking into another. Tau was defined initially in terms of comparing two rank orderings (tau *a*); it was then extended to R-analysis of square cross-tabulations

(tau *b*), and to cross-tabulations with unequal numbers of rows and columns (tau *c*). Tau measures reduce to phi (q.v. *infra*) for the 2×2 table and lie in the same range $(-1, +1)$ as rho. But there are some difficulties in interpreting tau when it is zero, and in stating the conditions under which the maximum is attained in the case of non-square tables.

Wilson's e: Table 2.2(4); Wilson (1974)
This coefficient resembles the tau family in that it is a test for strict monotonicity, but whereas the denominator of tau_b (which it most closely resembles) is $(\sqrt{\{C + D + T_x\}}\sqrt{\{C + D + T_y\}})$ the denominator of *e* is the much simpler quantity, $(C + D + T_y + T_x)$. Whilst *e* can never be greater than tau_b, its main property is its greater sensitivity to the extent to which data cluster round the main diagonal of a cross tabulation.

The relationship between these properly ordinal measures of association has been investigated—appropriately enough using MDS—by Maimon (1978), who shows the importance of the strict *vs* weak monotonicity distinction and the number of separate categories of the variables involved in accounting for differences between the measures.

2.2.3 Nominal level measures
A nominal scale consists simply of the division of a set of objects into a set of mutually exclusive and exhaustive categories, technically referred to as a partition. Three things are relevant to the analysis of such data:

(i)   *how many categories there are*—i.e. the 'fineness' of the partition, or the degree of discrimination, from the simplest dichotomy (2-state) to the multiple-state polytomy;
(ii) *how the objects are distributed over the categories*—i.e. the 'shape' of the partition, reflecting how common or how rare the occurrence of frequency of each category is;
(iii) *what the composition of the categories is*—i.e. the 'content' of the partition, indicating which particular objects occur within a category.

Partitions are rarely scaled as they stand. More typically, they are compared two at a time, and some measure of association is defined to summarise their (dis)similarity. In the case of R-analysis, two nominal level *variables* will be compared by means of a 2-way contingency table, where the rows represent the categories of one partition (say, sex), the columns represent the categories of the other partition (say, political affiliation), and the entries in the table consist of the number of subjects who fall in both the row category and in the column category. A large number of measures of association exist for assessing the similarity between the two variables, based upon the information in such tables. Many such measures are suitable for analysis by MDS, and are discussed below.

A particularly interesting special case occurs when the variables are dichotomies, where the contingency table is 2×2. Often the 'variables' in this case represent the presence or absence of some property, and the researcher is interested in the extent to which the two properties occur together (for instance, to what extent do two species of plant tend to grow in close proximity in a number of sites? Or, do two items in an attitude test evoke the same response in a sample of subjects? Or, do two coders agree in their identification of a theme in respondents' answers to a questionnaire?). Sometimes it will be sufficient simply to count how often the two properties occur together ('co-occurrence data', discussed in section 2.2.3.3), but in other cases the extent of disagreement will also be of interest, and a 'matching coefficient' will be necessary to express the overall similarity of the two dichotomies (see section 2.2.3.2).

In Q-analysis of nominal data, attention is focussed principally on comparing the structure of the individual partitions, taken two at a time. For instance, suppose a sample of subjects has been asked to sort a set of six objects into classes or categories of their own choosing, and the researcher wishes to examine how similar her subjects' classifications are. An important step will be to form a contingency table (as in R-analysis) but one which has the first subject's categories as the rows, and the second subject's categories as columns. The entries in the table will be the objects which are common to both the row category and the column category.

As in R-analysis, a number of measures exist for summarising the (dis)similarity between the two partitions, and these are discussed in section 2.2.3.4. We have written as if each partition in the Q-analysis comes from a separate subject. There is in fact no reason why the partitions should have been produced by individuals at all. Equally well, the subjects could be different times, occasions, methods, locations etc. (e.g., how reliable is a particular subject's classification system over a number of retests; or in different circumstances; or with different interviewers? How similar are census classifications of occupations in different countries, or over a number of revisions within the same country? How similar are the psychiatric diagnoses of a set of patients by doctors who have been trained in different traditions?)

2.2.3.1   Chi-square based measures
The most commonly-used family of measures compares the observed frequency of objects in each category with that which would be expected by chance. In the case of a cross-tabulation of two variables, the number expected by chance to be in class i of variable X and class j of variable Y (f_{ij}) is defined by statistical independence

$$f_{ij} = f_i f_j / N$$

where f_i is the total number in class i, f_j is the total in class j, and N is the total number of cases. Put in terms of proportions, independence is defined more simply as:

$$P_{ij} = P_i P_j$$

(or its equivalent $P_{ij} - P_i P_j = 0$).

The chi-square coefficient itself is a much-used test for statistical independence. But the value of chi-square is proportional to the number of cases, so it cannot serve as a measure of association for comparing groups of different sizes. Several attempts have been made to devise a measure of association based upon chi-square which will vary between 0 and 1 (the 'direction' or sign $(+, -)$ of a relationship is meaningless in the case of nominal data, since the classes may be arranged in any order), and will not depend on the number of subjects or the number of categories. The attempts to norm chi-square have not been entirely successful, but the measures presented in Table 2.3 are used fairly frequently in scaling analysis.

(χ^2 is the chi-square value; N is the total number of objects;
r is the number of rows in the 2-way table;
c is the number of columns, MIN is the smaller of r and c).

2.2.3.2   *Measures for dichotomies*
Dichotomous variables simply differentiate the presence and the absence of an attribute. Paradoxically, a dichotomy can be considered as being a nominal, ordinal or interval level variable:

(i)   The categories of a nominal level variable can always be converted into a set of dichotomies (thus the 4-fold religious categorisation 'Protestant, Catholic, Jew, other' can be converted into three* dichotomies: Protestant/not; Catholic/not; Jew/not.

(ii)   Presence/absence can be thought of as a particularly simple ordering.

(iii)   Since there is only a single difference (presence/absence) the numbers (1. 0) (or any linear transformation of them) can represent the two states quite legitimately, and indeed it is a common practice in regression and related linear models to follow this convention, calling them 'dummy variables'.

In constructing measures of association between two dichotomous variables, attention has been concentrated chiefly upon the question of 'matching'. This can best be illustrated by inspecting the 2 × 2 frequency table:

		Property Y		
		Yes	No	
Property X	Yes	a	b	$(a + b)$
	No	c	d	$(c + d)$
		$(a + c)$	$(b + d)$	$N(= a + b + c + d)$

Cells a and d represent positive and negative matches respectively; 'a' gives the number of individuals or objects who possess both property X and property Y (the positive matches) 'd' signifies those who possess neither X nor Y (the negative matches).

Cells b and c represent mismatches, individuals who possess one, but not the other property.

The large number of measures of association differ in large part in terms of (i) whether the 'negative matches' should enter into the assessment of similarity (i.e. are those who do not have either attribute even relevant in comparing properties?) and (ii) what weight should the matches and the mismatches have in defining the degree of similarity?

*N.B.   One category must be omitted in converting nominal scale to dichotomies, since the response to the omitted category is perfectly predictable from knowing the response on the others. For instance, knowing that someone is not Protestant, not other and not Jewish implies that the person is Catholic since the categorisation must be exclusive and exhaustive to qualify on a nominal scale.

Name	Formula	Limits			Advantages	Disadvantages	Reduces to
		Maximum	Zero	Minimum			
CHI SQUARE (χ^2)	$\sum \frac{(f_o - f_e)^2}{f_e}$	$N(\text{MIN}-1)$	Statistical Independence (SI)	0	Zero in case of independence. Serves as basis for other measures	Non normed. Dependent on N	—
PHI (φ)	$\sqrt{\left(\frac{\chi^2}{N}\right)}$	$\leqslant (\text{MIN}-1)$	SI	0	Normed, Independent of N. Reaches maximum for 2×2 tables	Can exceed 1 as maximum when $\text{MIN} > 1$	PM Correlation for 2×2 table
PEARSON'S CONTINGENCY COEFFICIENT C	$\sqrt{\left(\frac{\chi^2}{\chi^2 + N}\right)}$	Depends on r and c	SI	0	Normed, Independent of N	Can neither exceed, nor reach, 1 as maximum	—
TSCHUPROW'S COEFFICIENT T	$\sqrt{\left(\frac{\chi^2}{N\sqrt{(r-1)(c-1)}}\right)}$	Depends on r and c	SI	0	Normed, Independent on N. Reaches maximum (1) for square tables	Cannot reach 1 as maximum in non-square tables	Phi (in 2×2 case)
CRAMER'S COEFFICIENT V	$\sqrt{\left(\frac{\chi^2}{N(\text{MIN}-1)}\right)}$	Unity	SI	0	Normed, Independent of N. Reaches maximum, even for non-square tables. Independent of number of rows and columns		T, if $r = c$ in 2×2, and $2 \times k$ cases

Table 2.3   Chi-square based measures of association

The superabundance of measures of dichotomous association is due in large part to their importance in numerical taxonomy, where the basic crucial operation is often the comparison of pairs of OTUs (operational taxonomic units) or individual organisms who share a number of attributes to a greater or lesser extent (Sokal and Sneath 1963; Sneath and Sokal 1973) and in the social sciences, where dichotomous variables abound, though here use has generally only been made of a very restricted number of coefficients (but see Lazarsfeld and Henry 1968 and Rasch 1960 for a wide variety of models based upon the covariation of dichotomous data).

Sokal and Sneath (1963) present a particularly useful classification of measures of (dis)similarity for dichotomous data, which in simplified form provides the basis for Table 2.4. The measures all take the form of a ratio between the number of 'matches' (numerator) and the elements considered to be the relevant reference set (the denominator). They differ in two major respects:

(i)   *How 'matching' is to be defined in the numerator*—in particular whether the negative matches (cell d) are to be excluded (I), included (II), or whether the numerator should take into account matched *and* unmatched pairs (III).

(ii)   *What weight is to be given to the relative preponderance of matched and unmatched pairs.*   Here there is greater variety, with quantities such as marginal totals (e and h) and the sum of cross products (g) entering the definition of the denominator.

Brief comments on the properties of some of these measures are given below:

Measure number:

1   Represents the conditional probability that a pair of objects will both have a randomly chosen variable, and is one of the simplest and longest used coefficients, which excludes negative matches.

2   A curious measure which by implication treats negative matches as generically different from positive matches.

3   The 'simple matching coefficient', which includes negative matches.

7   This measure was defined originally for polytomies, and allows missing data.

8 and 9   Despite their apparently simple interpretation, both these measures have the unfortunate property of being normed between 0 and *infinity*, unlike the other measures which all have an upper limit of unity.

14   Unlike the other measures, this bases association on the preponderance of matches over mismatches.

15   Used extensively in social science data analysis, and based like phi on the determinant of the table. It shares the unfortunate property with phi that if no negative matches occur (i.e. if $d = 0$), then an association of zero results.

16   Phi has been extensively discussed, and is widely used as the dichotomous equivalent of the Pearsonian PM correlation coefficient.

NUMERATOR

DENOMINATOR†	(I) EXCLUDES negative matches (d)	(II) INCLUDES negative matches (d)
(a) m and u equally weighted	1 Jaccard $a/(a + b + c)$ 2 Russell and Rao $a/(a + b + c + d)$	3 Sokal $(a + d)/(a + b + c + d)$
(b) m given twice weight of u	4 Dice $2a/(2a + b + c)$	5 (s.n.)* $2(a + d)/(2(a + d) + b + c)$
(c) u given twice weight of m	6 (s.n.)* $a/(a + 2(b + c))$	7 Rogers and Tanimoto $(a + d)/(a + d + 2(b + c))$
(d) u only	8 Kulczynski $a/(b + c)$	9 (s.n.)* $(a + d)/(b + c)$
(e) marginal totals	10 Kulczynski $\frac{1}{2}\{a/(a + c) + a/(a + b)\}$ 12 Ochiai $a/\sqrt{\{(a + c)(a + b)\}}$	11 (s.n.)* $\frac{1}{4}\{a/(a + c) + a/(a + b) + d/(b + d) + d/(c + d)\}$ 13 (s.n.)* $ad/\sqrt{\{(a + c)(a + b)(b + d)(c + d)\}}$

NUMERATOR

DENOMINATOR†	(III) BALANCE of matched and unmatched pairs
(f) m and u equally weighted	14 Hamann $(a + d) - (b + c)/(a + b + c + d)$
(g) sum of cross products	15 Yule's Q $(ad - bc)/(ad + bc)$
(h) marginal totals	16 Pearson's Phi $(ad - bc)/\sqrt{\{(a + c)(a + b)(b + d)(c + d)\}}$

*sine nomine (unnamed coefficient)
†Note m signifies matched pairs (a and d); u signifies unmatched pairs (b and c).

Table 2.4 Measures of association between dichotomies (based upon Sokal and Sneath 1963, pp. 125 et seq.)

2.2.3.3   Co-occurrence measures

Co-occurrence measures are all based upon the 'abundance matrix' (Kendall 1971a, p. 219) whose entries s_{jk} assess the similarity between objects j and k in terms of the frequency with which objects j and k occur in (or are allocated to) the same category. Miller (1969, p. 171 et seq.) has shown that the simple measure: $(N - s_{jk})$, where N is the total number of subjects or partitions, obeys the axioms of a distance metric (see A2.1.1) and is frequently used in MDS as the basic dissimilarity measure between objects. In some applications, the user may be advised to modify this simple measure by taking into account the size of the category in which each pair of objects occurs. Burton (1975, and see Coxon and Jones 1979, U2.10) has defined a family of four co-occurrence measures which differ in this respect:

M1   Each individual co-occurrence contributes equally to the overall measure (the basic measure).

M2   Each co-occurrence is weighted by the size of the category in which it occurs (thus emphasising gross discriminations).

M3   Each co-occurrence is weighted *inversely* by the size of the category (emphasizing fine discriminations).

M4 An information theoretic measure which also emphasises fine discriminations, and in addition, takes into account the number of times in which a pair of objects are sorted into *different* groups.

(These measures are defined in Appendix A.2). Burton (1975) examined how well measures M1, M2 and M4 can be scaled using the basic MDS model, and the effect which the differences between the measures have upon the structure of the final configuration of points. As in other applications (Burton and Romney 1975; Coxon and Jones 1979, U2.11), M2 (which emphasizes gross distinctions) gives the best fit, but is liable to collapse points into large clusters. M4 is less readily representable by the basic MDS model, but is probably the most satisfactory measure for MDS analysis, due both to its greater resistance to 'degeneracy' (i.e. the tendency to collapse points into clusters, ignoring significant information in order to get a better fit) and because, like M3, it attempts to take into account both the tendency for objects to occur together in some partitions *and* for them to be separated in others. In the sense that it balances concordant and discordant pairs, M4 resembles the ordinal measures of association.

2.2.3.4   The index of dissimilarity between distributions

Very commonly, data analysts wish to compare two distributions of a categorical variable—such as the incidence of a number of diseases in two countries, or in two social classes. A particularly simple measure of how (dis)similar two distributions are is provided by the 'index of dissimilarity' (see Blau and Duncan 1967, p. 43 et seq.) which is based simply on the percentage difference for each category and illustrated below.

Each distribution is first converted into percentage form (for comparability). Then for each category one calculates the difference (e.g., $12.3 - 25.8 = -13.5$ for category A). The absolute differences—ignoring the sign—are then added to form the basic index (here, 60.2). Clearly, if the percentage distributions had been

Category	Group: I	Group: II	% diff.	Absolute % diff. (AD)
A	12.3%	25.8%	−13.5	13.5
B	5.8	13.2	−7.4	7.4
C	54.2	40.3	13.9	13.9
D	22.4	6.2	16.2	16.2
E	5.3	14.5	−9.2	9.2
Total	100.0	100.0	0.0	60.2
N =	532	821		

INDEX OF DISSIMILARITY (ID) BETWEEN GROUPS I AND II $= \Sigma \, AD/2 = 30.1$

identical the value of the index would be 0 indicating 'zero dissimilarity'. If the index is halved, forming the index of dissimilarity (ID), it has a particularly simple interpretation: it represents the percentage of cases which would have to be moved between categories in order to change one distribution into the other—a notion encountered already in ordinal and other nominal measures of association. In this example, 30.1 per cent of cases would need to change categories if the distributions were to become identical. The index of dissimilarity is a distance measure: it is zero only if the distributions are identical, it is symmetric, and for comparisons between three distributions it obeys the triangle inequality. Moreover, its value is unaffected by re-arrangement of the order of the categories and is therefore appropriate to nominal data.

The ID measure has been frequently used to analyse 'flow data'—for example, mobility between occupational groups, migration between regions, input/output analysis between economic sectors, volume of diplomatic correspondence between countries, settlement of plant species in different sites. In many cases the flow is asymmetric in the sense that not as many objects move from category a to category b as do from category b to category a. A common way of analysing these data is first to convert the raw frequency ('turnover') table into a row-percentaged table (for assessing 'outflow' movement from a given row category into the column categories) and secondly into a column-percentaged table (for assessing 'inflow' into a given column category from the row categories). This is illustrated in Table 2.5 for some (fictional) migration data between four cities.

The index of dissimilarity can now be used to summarise these rather complex flow data. In the case of outflow each pair of rows is compared by calculating the index (how dissimilar are cities j and k in terms of the destinations of their inhabitants?)—in this instance, cities B and C are most alike in their pattern of outmigration (ID = 7.8), and cities A and D are least alike (ID = 54.9). For inflow data, each pair of columns is compared (how dissimilar are cities j and k in terms of their in-migration or recruitment?). The values of the ID coefficients for both outflow and inflow are given in the lowest table.

The ID coefficient is used extensively as a prelude to MDS analysis (Blau and

I   RAW AND CONDITIONAL PERCENTAGED CROSS TABULATION

(a)   *Raw frequencies*

		To city				
		A	B	C	D	Total
From city	A	58	22	41	19	140
	B	30	38	14	23	105
	C	25	44	19	22	110
	D	7	51	12	51	121
	Total	120	155	86	115	$N = 476$

(b)   *Row-percentaged data ('outflow')*

	A	B	C	D	Total
A	41.4	15.7	29.3	13.6	100.0 per cent
B	28.6	36.2	13.3	21.9	100.0
C	22.7	40.0	17.3	20.0	100.0
D	5.8	42.1	10.0	42.1	100.0

(c)   *Column-percentaged data ('inflow')*

	A	B	C	D	Total
A	48.4	14.2	47.6	16.6	
B	25.0	24.5	16.3	20.0	
C	20.8	28.4	22.1	19.1	
D	5.8	32.9	14.0	44.3	
Total	100.0	100.0	100.0	100.0	per cent

II   INDEX OF DISSIMILARITY VALUES

Above diagonal: based on row percentages ('outflow')
Below diagonal: based on column percentages ('inflow')

	A	B	C	D	
A	—	28.7	30.7	54.9	
B	34.7	—	7.8	27.0	
C	9.5	33.4	—	24.2	
D	38.5	13.8	34.0	—	

Table 2.5   *Calculation of index of dissimilarity*

Duncan 1967, p. 67 et seq.; Macdonald 1972, p. 213 et seq.), especially for asymmetric data of this sort.* It is further discussed in section 5.1.1.1 below.

2.2.3.5   Similarity between pairs of partitions
Although Q-analysis of sortings and partitions is a fairly recent development in MDS analysis, it is receiving increasing attention, especially following the

---

*Blau and Duncan further discuss the possibility of leaving out the diagonal elements corresponding to the 'stayers' in calculating an index of dissimilarity on the grounds that it is intended to assess the dispersion or flow, not the stability, between distributions.

important methodological work of Boorman and Arabie (Arabie and Boorman 1973). Although the literature is at a fairly technical level, the basic concepts are surprisingly simple. The idea is to provide a measure of dissimilarity between any two partitions by examining how many moves it will take to change one partition into the other. As an example, take the following three partitions:

$$S = \{A \quad | BDC | FE\}$$
$$T = \{AD \ | \quad BC | FE\}$$
$$U = \{ACF| \quad B | E | D\}$$

(The order of the categories, and of the objects within a category, is arbitrary). It is fully obvious that S and T are very similar, both in the number and in the composition of their categories, whilst they both differ substantially from U. The question is how different they are, and a quantitative answer depends entirely upon how a 'move' or 'change' is defined. If the move consists of single elements, then moving D out of (BDC) and into conjunction with A will turn S into T. However, if we wish to preserve information about what objects are linked together in the same category, then we shall at least need to move three *pairs* of elements—(BD), (CD), and (AD)—to change S into T. And so on: the definition of a move depends upon how much structure one wishes to preserve intact in moving from one partition to another.

To show the usefulness and versatility of the Boorman-Arabie group of measures, the pairwise ('Pairbonds') dissimilarity measure between S and T will serve as an example. In all, there are 15 possible pair-linkages between the 6 elements: $\{A, B, C, D, E, F\}$. When these are enumerated and illustrated as a Venn Diagram, (Figure 2.1a) it can be seen that S and T *agree* that (BC) and (EF) go together and that a further 10 pairs do not go together. The disagreement between S and T is limited to three pairs: the pairs (BD) and (CD) (which occur in S but not in T) and the pair (AD) (which occurs in T but not in S). These three pairs together define the distance, or Pairbonds dissimilarity measure, between S and T: $d_{\text{pairbonds}}(S, T) = 3$. The representations of the Pairbonds measure in Figure 2.1 emphasise some important parallels with measures we have already discussed. First, Pairbonds is clearly another matching measure between two partitions, but is one which counts the *mismatches* in the pairs involved (i.e. the count $(b + c)$ in Figure 2.1c). Secondly, the Venn Diagram makes it clear that Pairbonds is in set-theoretic terms the symmetric difference of the pairs involved in the partitions S and T. Indeed, Flament (1963; pp. 14–17) and Restle (1959) before him, discuss precisely this measure, prove that it is a metric, and Flament shows that it may usefully be employed to compare the dissimilarity of two graphs such as communication and friendship networks, and the dissimilarity of any two binary $(0, 1)$ relations. (In this context, Pairbonds provides a natural Q-analysis comparison to the R-analysis measures of pairwise co-occurrence discussed in the previous section). Thirdly the formula for the Pairbonds dissimilarity measure, like many others discussed by Boorman, has a familiar form, akin to the cosine rule discussed in Appendix A2.1:

$$d(S, T) = m(S) + m(T) - 2m(S \text{ and } T)$$

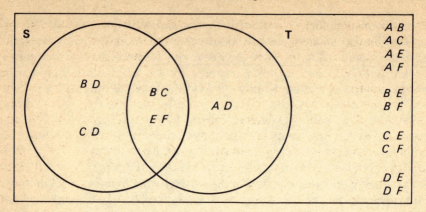

(a) Venn diagram of pairs

	T:		
Partition	*AD*	*BC*	*EF*
A	*A*	–	–
S: *BDC*	*D*	*BC*	–
FE	–	–	*FE*

(b) Intersection table of objects

Pairs in T

Pairs in S		Y	N
	Y	2	2
	N	1	10

$N = 15$

(c) Matching table of pairs

Figure 2.1 *Pairbonds measure of dissimilarity between partitions*

In the case of Pairbonds, the measure m consists simply of the number of pairs involved:

$$m(S) = 4 \quad \text{i.e.} \quad (BD), (CD), (BC), (EF)$$
$$m(T) = 3 \quad \text{i.e.} \quad (BC), (EF), (AD)$$

The table of the intersection between S and T, which is similar to the contingency table in R-analysis, is given in Figure 2.1b. It is produced simply by cross-classifying the two partitions, and writing in the cell (i, j) the elements which are in both category i of S and category j of T. (Taken together, the entries in the table incidentally comprise a partition which is 'finer' than either S or T, in the sense that both S and T can be built up from it). There are two pairs, namely BC and FE, which occur in the intersection table, hence

$$m(S \text{ and } T) = 2$$

Thus: Pairbonds: $d(S, T) = m(S) + m(T) - 2m(S \text{ and } T)$
$$= 4 \quad + 3 \quad - 2 \times 2$$
$$= 3$$

A further eleven measures of dissimilarity between partitions are defined by Arabie and Boorman (1973) in a similar manner. They also investigate in a very instructive manner the behaviour of these measures when submitted to MDS analysis. We shall examine some of their conclusions in Chapter 4. Examples of the use of the measures are provided in Arabie and Boorman (*op. cit*) and in Coxon and Jones (1979).

Unfortunately, partitions do not provide any information on the *relationship* between the categories, but it is a simple matter to supplement nominal information if the data are obtained directly from the subjects, either by asking them to place the categories in order (thus producing a weak ordering), or to continue merging categories in terms of their relative similarity, which produces a hierarchical clustering of the objects (see above). Boorman and Olivier (1973) have used the partition measures as a basis for constructing measures of similarity between hierarchies of this sort, and these are employed in Coxon and Jones (1978b) to scale differences between subjective hierarchies.

2.3 Summary

Two main criteria have been used to distinguish the profusion of measures of association appropriate for scaling analysis:

(i)   direct *vs* derived (aggregate) measures;
(ii) the level of measurement of the data or of the coefficients.

The distinction between direct and aggregate data is very important, and the two types of data differ in terms of information loss and the form of data produced. Indirect measures are summaries calculated from original data, and are produced by aggregating over one facet of the data (over subjects in the case of R-analysis, and over objects in Q-analysis). By contrast, direct measures preserve individual data intact and make it possible, at least in principle, to detect systematic individual differences. Secondly, aggregate measures take the form of a coefficient of (dis)similarity between each pair of objects (or subjects) and are almost without exception 'well-behaved' measures, obeying the triangle inequality. Since aggregate measures produce a square symmetric matrix of (dis)similarity coefficients, they can all be analysed using the basic MDS model. By contrast, the properties of direct measures are *not* known in advance, and it is quite likely that, as they stand, some of the data may be inconsistent with any numerical representation. On the other hand, most direct forms of data can be analysed by specifically designed programs, which allow the researcher to examine how well each set of individual data fits the overall configuration.

The question of the level of measurement of the data is also important, but is not always crucial. It is of course sensible for the researcher to choose a measure of association which matches as closely as possible the level of measurement of the variables concerned, and it is advisable to use more than one measure in order to assess the extent to which results are dependent upon the properties of particular measures. But one of the main uses of non-metric MDS is to see whether, whilst making very conservative claims for the level of measurement of the data, it is possible to find a legitimate transformation (re-scaling) which will yield a much higher level, better behaved, set of values.

Normally, the most general rule is to preserve as much information from the original data as possible when choosing data for input to MDS. In the case of direct data, it is important to choose a program which matches the type of data as closely as possible. For aggregate measures, it is important to choose a dis(similarity) measure whose properties are well understood, and which match the level of measurement of the data as closely as possible.

Level of Measurement	*Type of Data* Direct (Method of data collection)	Derived/Aggregate (Measure of (dis)similarity)
INTERVAL	Ratings, flow rates, latencies	Product moment measures, (especially covariance, correlation)
ORDERED METRIC	Tetrads Triads	
ORDINAL: *Strict*	Rankings	Tau measures, and Wilson's *e*
Weak	Rankings, with ties	Gamma
HIERARCHICAL	Hierarchies (rooted trees)	Boorman-Olivier family of measures
Polytomies NOMINAL *Dichotomies*	Partitions Pair comparisons	Chi-square based measures (esp. Cramer's *V*) Co-occurrence measures (esp. Burton's *Z*) Boorman-Arabie family of measures Matching coefficients Symmetric difference

Table 2.6   *Types of (dis)similarity measures for scaling*

APPENDIX A2.1   DISTANCE MEASURES AND SCALAR PRODUCTS

A.2.1.1   Distance measures

The correspondence between dissimilarity and distance has been analysed rigorously in mathematics in terms of the notion of a 'metric' or general distance measure (of which Euclidean distance is a special case). Most measures of association used in statistics and data analysis, for example, satisfy the requirements of such a general distance measure, though this does not necessarily mean that they can be directly represented in a Euclidean space. Indeed, one purpose of non-metric MDS is to see whether we can *re-scale* data into a set of quantities which are capable of such a representation.

There are two basic properties (or axioms) which a measure must satisfy to count as a metric, and a further one is also normally required. These are listed below. ($d(A, B)$ should be read: the distance between points A and B):

Properties of a Distance Measure

(i)   *Non-negativity and equivalence*
$$d(A, B) \geqslant 0 \quad \text{for all points } A, B$$
and
$$d(A, B) = 0 \quad \text{if and only if } A \text{ coincides with } B$$

(ii) *Symmetry*
$$d(A, B) = d(B, A) \quad \text{for all points } A, B$$

(iii) *Triangle inequality*
$$d(A, C) \leqslant d(A, B) + d(B, C) \quad \text{for all points } A, B, C$$

A measure that satisfies properties (i) and (ii) but not (iii) is often termed a 'semi-metric'; one that satisfies all three axioms is referred to as a 'metric'. Several comments on these properties are appropriate.

Non-negativity might seem to exclude covariances, correlations etc., which can take on negative values. This difficulty can be overcome fairly easily.

Symmetry might seem to exclude asymmetric dependence measures, such as regression coefficients, although asymmetry *can* be represented in a spatial manner (see 5.1.1.1).

Triangle inequality is the most restrictive axiom, and can best be illustrated by the fact that in Euclidean space a point B must either lie *on* the line AC, in which case $d(A, C) = d(A, B) + d(B, C)$, or else it must lie *off* the line AC, in which case the sum: $d(A, B) + d(B, C)$ must exceed $d(A, C)$.

The Triangle Inequality Axiom

(a) *Triangle equality*   (B lies on line AC)
$$d(A, C) = d(A, B) + d(B, C)$$

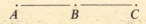

(b) *Triangle strict inequality*   (B lies off line AC)
$$d(A, C) < d(A, B) + d(B, C)$$

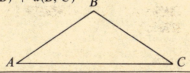

This axiom clearly excludes the possibility that
$$d(A, C) > d(A, B) + d(B, C)$$

A large number of measures used in empirical research satisfy these three axioms but it is by no means clear simply by inspection whether they do or do not.

A2.1.1.1   Euclidean distance

The most common way of representing dissimilarity in MDS is in terms of Euclidean distance, which involves a surprisingly restrictive set of further assumptions. The user should be aware of what these are. The mathematical definition of Euclidean distance is as follows:

$$d_{jk} = \sqrt{\left\{\sum_a (x_{ja} - x_{ka})^2\right\}}$$

where x_{ja} refers to the co-ordinate of point j on dimension a.

The formal and substantive assumptions of Euclidean distance have been investigated from a measurement theory viewpoint by Beals et al. (1968). Four definitional characteristics of Euclidean distance are of special relevance in elucidating the assumptions implicit in the use of the distance model (Tversky and Krantz 1970, p. 4):

(i) *Decomposability* The distance between the objects can be decomposed into a contribution from each of the dimensions (a) of the space.

(ii) *Intra-dimensional subtractivity* $(x_{ja} - x_{ka})$
Each contribution to the distance between two points is composed of the difference in scale values within each dimension.

(iii) *Inter-dimensional additivity* (the summation over a)
The distance measure combines these differences additively from each dimension.

(iv) *Metric* (the squaring of the differences)
All the differences in (ii) are transformed by the same power-function.

As the authors show, it is possible to state the empirical conditions which data must satisfy if at least the first three of these characteristics are to be justifiable. Users should note that there is no guarantee that a given set of ordinal similarity data can be embedded in a metric space and that the metric and dimensional assumptions of the distance model are quite distinct; it is quite possible that some data will satisfy the latter but not the former set of assumptions.

A2.1.1.2 Minkowski metrics
The Euclidean metric is a special case of a more general family of distance measures, referred to under as Minkowski r (or L_p or power) metrics of the form:

$$d_{jk}^{(r)} = \sqrt[r]{\sum_{a=1} |x_{ja} - x_{ka}|^r}$$

Each value of r ($\geqslant 1$) substituted in this formula defines a distinct metric, all of which obey the triangle inequality. Clearly, if $r = 2$ then the Euclidean metric results. Other values used in scaling include the city-block (or 'Manhattan' or 'taxi-cab') metric (where $r = 1$), and the dominance metric (where $r = \infty$). The properties of these metrics and their applications in MDS are discussed under 5.3.3.2. Carroll and Wish (1974a, p. 412 et seq.) give an extended treatment and also consider the case where $r < 1$, and other metric families such as Riemannian metrics of constant curvature which have also been used in MDS.

A.2.1.2 Distance and vector representation of data
Consider the following data matrix:

$$\mathbf{X} = \begin{bmatrix} 5 & 1 \\ 3 & 2 \\ 1 & 4 \end{bmatrix}$$

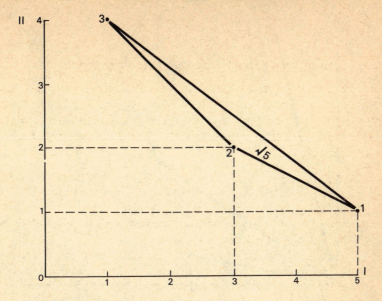

Distance between variables 1 and 2

$$d(1,2) = \sqrt{(5-3)^2 + (1-2)^2}$$
$$= \sqrt{(4+1)}$$
$$= 2.236$$

Matrix of distances

Variable	1	2	3
1	0	$\sqrt{5}$	$\sqrt{25}$
2	$\sqrt{5}$	0	$\sqrt{8}$
3	$\sqrt{25}$	$\sqrt{8}$	0

Figure A2.1   *Euclidean distances between three variables*

The information in this matrix can be thought of *either* as giving the co-ordinates for locating two (column) elements in three (row) dimensional space, *or* for locating three (row) elements in two (column) dimensional space. For simplicity of exposition we shall assume that it locates three variables describing two subjects, and that we wish to assess the similarity of the variables (R-analysis).

If a distance measure of similarity is required, then this can be calculated using the Euclidean distance formula illustrated in Figure A2.1.

Two operations often performed on raw data scores are:

(i) *centring*—creating the 'deviate score' by removing the overall effect of a variable by subtracting the mean.

(ii) *standardising*—creating the 'normal score' by dividing the deviate score by the standard deviation of the variable, thus reducing all variables to a common unit of measurement.

Centring the variables has the geometric effect of removing the origin of the space

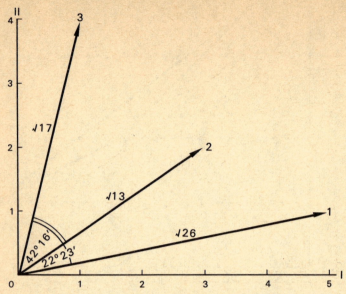

Figure A2.2   *Vector representation of similarity*

(0, 0) to the centroid (centre of gravity) of the points defined by the means of the variables, i.e. to (3, 2) in this case, but it does not affect the distances in any way. By contrast, standardising *does* affect distances since the axes are now differentially stretched to a common unit.

The vector representation of the same data is given in Figure A2.2. The scalar products measure of similarity between the variables is produced by forming the (major) product moment matrix, $\mathbf{A} = \mathbf{XX'}$, which in this case is:

$$\begin{bmatrix} 5 & 1 \\ 3 & 2 \\ 1 & 4 \end{bmatrix} \begin{bmatrix} 5 & 3 & 1 \\ 1 & 2 & 4 \end{bmatrix} = \begin{bmatrix} 26 & 17 & 9 \\ 17 & 13 & 11 \\ 9 & 11 & 17 \end{bmatrix}$$

$$\quad \mathbf{X} \qquad\qquad \mathbf{X'} \qquad = \qquad \mathbf{A}$$

This is often referred to as the matrix of 'crude sums of squares and cross products' (CSSCP) in the multivariate analysis literature. The *minor* product matrix $\mathbf{X'X}$ (of order two) provides the scalar products between the two individuals, aggregated over the variables). The entries in the product moment matrix are readily interpretable. The diagonal entry, a_{ii}, gives the squared length of the vector draw from the origin to the point i (call it l_i^2). The symmetric off-diagonal elements a_{ij} give the scalar product between vector i and j, which is related to the angular separation between the vectors. Explicitly it is the product of the length of each vector and of the cosine of the angle separating them: (see van der Geer 1971, pp. 19–21)

$$a_{ij} = l_i l_j \cos \theta_{ij}$$

e.g. in the case of variables 1 and 2:

$$a_{12} = \sqrt{26}\sqrt{13} \cos (22°23') \simeq (5.100)(3.606)(0.924)$$
$$= 17.000$$

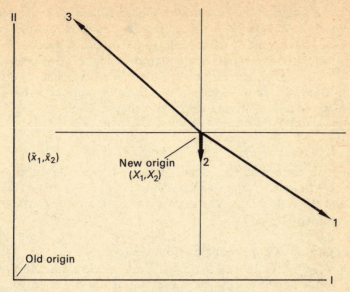

Figure A2.3 *Vector representation from centroid origin*

Although centring has no effect upon distances, it dramatically alters the measure of similarity based upon scalar products, since the lengths of the vectors and their angular separation is now assessed from a different origin. The effect of centring is illustrated in Figure A2.3. The product moment matrix **B** formed from the centred deviate matrix $\mathbf{X}_d$ is as follows:

$$\mathbf{B} = \mathbf{X}_d\mathbf{X}_d' = \begin{bmatrix} 2 & -\frac{4}{3} \\ 0 & -\frac{1}{3} \\ -2 & \frac{5}{3} \end{bmatrix} \begin{bmatrix} 2 & 0 & -2 \\ -\frac{4}{3} & -\frac{1}{3} & \frac{5}{3} \end{bmatrix}$$

$$= \begin{bmatrix} 5.78 & 0.44 & -6.22 \\ 0.44 & 0.11 & -0.56 \\ -6.22 & -0.56 & 6.78 \end{bmatrix}$$

Notice that the scalar products between 1 and 3 and between 2 and 3 are now negative. Obviously, centring variables does *not* leave vector separation measures unchanged. Consequently when vector or factor model solutions are presented in MDS the origin of the configuration is fixed, and may not be relocated at will. By contrast, the origin in Euclidean distance model configurations is arbitrary and may be moved.

There are two especially useful variants of the relation: $a_{ij} = l_i l_j \cos \theta_{ij}$, which apply when the variables are centred, and normalised.

(i) If the $\mathbf{X}_d\mathbf{X}_d'$ matrix is multiplied throughout by the constant $1/N$ (where N is the number of individuals), the resulting matrix consists of the dispersion matrix, Σ, which features centrally in multivariate analysis, whose diagonal elements σ_{ii} are the variances, and whose off-diagonal elements σ_{ij} are the covariances between the variables i and j. In this case, the relation is:

$$a_{ij} = \sigma_i \sigma_j \cos \theta_{ij} = \sigma_{ij} \quad \text{(covariance)}.$$

Again, both the length of the variables (in this case, their dispersion) and their angular separation contribute to the covariance measure, which is therefore sensitive to the scaling (measurement units) of the original variables.

(ii) If in addition the variables are standardised (to zero mean and unit variance $z_{ij} = (x_{ij} - \bar{X}_j)/\sigma_j$), then the dispersions all become unit length, which is tantamount to saying that the original variable scaling units are arbitrary. Then the form of the relation becomes especially simple. Since $\sigma_i = \sigma_j = 1$ for standardised variables, then it reduces to:

$$a_{ij} = (1)(1) \cos \theta_{ij} = \cos \theta_{ij} = r_{ij} \quad \text{(correlation)}.$$

Hence Pearson's correlation coefficient preserves *only* the angular separation *in normalised axes* (shrunk or expanded in order to equalise dispersions) between the variables, and the scale units of the original values in no way contribute to the measure of similarity.

A.2.1.3   Coverting scalar products into distances

Conversion of scalar products into Euclidean distances involves a simple application of the cosine rule, which states that in a non-right-angled triangle, $c^2 = a^2 + b^2 - 2ab \cos \theta$.

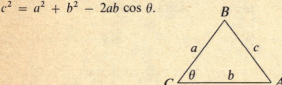

Thinking of CB and CA as vectors, θ as the angle separating them, and AB as the distance $d(A, B)$ corresponding to the angular separation, then the rule may be rewritten as:

$$
\begin{aligned}
d(A, B) &= l_a^2 + l_b^2 - 2l_a l_b \cos \theta_{ab} \\
&= l_a^2 + l_b^2 - 2 \times \text{(scalar product between } a \text{ and } b) \\
&= a_{ii} + a_{jj} - 2a_{ij} \quad \text{(in terms of the product moment matrix of scalar products)}
\end{aligned}
$$

For example, in Figure A2.2,

$$
\begin{aligned}
d_{12}^2 &= 26 + 13 - (2)(17) \\
&= 39 - 34 = 5 \\
d_{12} &= \sqrt{5}, \quad \text{which is the quantity calculated in Figure A2.1.}
\end{aligned}
$$

A special application of the cosine rule, and remembering that standardised variables are unit length, shows that correlations are (inversely) monotonic with distances.

The cosine rule:

$$d_{ij}^2 = l_i^2 + l_j^2 - 2l_i l_j \cos \theta_{ij}$$

Since

$$r_{ij} = \cos \theta_{ij},$$

then

$$d_{ij}^2 = 2 - 2r_{ij}$$

and

$$d_{ij} = \sqrt{(2 - 2r_{ij})} = \sqrt{2(1 - r_{ij})}.$$

Clearly, the relationship between distance and correlation is a decreasing one (in effect, $1 - r_{ij}$ forms the *dis*similarity coefficient), and it is *non*-linear (because of the square root). But there *is* a monotonic relationship between distance and correlation, including negative correlation values.

The conversion of Euclidean distances into scalar product form is a slightly more complicated matter, and is taken up in Appendix A5.2.

APPENDIX A2.2   CO-OCCURRENCE MEASURES OF SIMILARITY

An individual partition of a set of N objects or elements can be represented as a square symmetric $(0, 1)$ matrix $\mathbf{S}$ of order N, where $S_{ij} = 1$ if objects i and j occur in the same category of the partition, and $S_{ij} = 0$ otherwise.

Thus, the partition $I = \{3, 1 \,|\, 4, 2, 5\}$ can be represented by the co-occurrence or 'incidence' matrix:

$$\mathbf{S}^{(I)} = \begin{bmatrix} 1 & 0 & 1 & 0 & 0 \\ 0 & 1 & 0 & 1 & 1 \\ 1 & 0 & 1 & 0 & 0 \\ 0 & 1 & 0 & 1 & 1 \\ 0 & 1 & 0 & 1 & 1 \end{bmatrix}$$

Clearly, co-occurrence is a metric, obeying the triangle inequality since i cannot occur with j, and j occur with k without i also occurring with k.

When there is a set of r partitions, the $\mathbf{S}$ matrices are simply added together to form the aggregate co-occurrence matrix. It also represents a metric, since the sum of metrics is a metric. The four measures referred to in the text (2.2.3.3) and defined in Burton (1975) differ basically in how the size of the category is taken into account before the individual matrices are aggregated.

M1   (the basic measure, cf. Miller 1969, which is called F in Burton 1975)
Each co-occurrence contributes equally, so the aggregate matrix $\mathbf{S}$ is the simple sum of the $(1, 0)$ individual matrices.

M2   (called G in Burton 1975)
The entries in the individual co-occurrence matrix $\mathbf{S}^{(I)}$ are *the number of elements in the category* from which the pair is drawn. In this case, the individual co-occurrence matrix corresponding to partition I would be:

$$\mathbf{S}^{(I)} = \begin{bmatrix} 2 & 0 & 2 & 0 & 0 \\ 0 & 3 & 0 & 3 & 3 \\ 2 & 0 & 2 & 0 & 0 \\ 0 & 3 & 0 & 3 & 3 \\ 0 & 3 & 0 & 3 & 3 \end{bmatrix}$$

(N.B.   diagonal elements are ignored)

Hence, the larger the category in which a pair of objects occur, the higher their similarity is considered to be.

M3   The entries in $S^{(I)}$ are the *reciprocal of the number* of elements in the category from which the pair is drawn. In this case,

$$\begin{bmatrix} (-) & 0 & \frac{1}{2} & 0 & 0 \\ 0 & (-) & 0 & \frac{1}{3} & \frac{1}{3} \\ \frac{1}{2} & 0 & (-) & 0 & 0 \\ 0 & \frac{1}{3} & 0 & (-) & \frac{1}{3} \\ 0 & \frac{1}{3} & 0 & \frac{1}{3} & (-) \end{bmatrix}$$

Clearly, this measure *deflates* similarity by the size of category, on the reasoning that the more unusual co-occurrences, or the more fine discriminations denote greater similarity.

M4   (called *Z* in Burton 1975)
This information-theoretic measure, which is akin to *M3* in emphasising the similarity of pairs from small categories, is based upon the 'surprisal value' of each category. This is defined in terms of

(i)   the probability that two objects *j* and *k* will be found in the *same* category, *a*.

$$p_a^{(I)} = n_a(n_a - 1)/n$$

(where n_a is the number of objects in category *a*, and *n* is the total number of pairs) and
(ii)   the probability that *j* and *k* will be found in,

$$Q^{(I)} = 1 - \sum_a p_a^{(I)}$$

The contribution which each pair of objects (*j*, *k*) makes is defined as its surprisal value.

$$-\log_2 (p_a^{(I)}) \text{ if } j \text{ and } k \text{ are in the same group}$$
and $$-\log_2 (Q^{(I)}) \text{ if } j \text{ and } k \text{ are in different groups}$$

Since surprisal is negatively related to the size of the group, *M4* also emphasizes finer discriminations, but (unlike *M3*) makes explicit allowance for pairs occurring in different categories. In the present case, since the probability of two objects being in the same category is 0.1 for category 1, and 0.3 for category 2, and the probability for being in different categories is 0.6, the matrix of similarity (surprisal values) is:

$$\begin{bmatrix} - & 0.74 & 3.32 & 0.74 & 0.74 \\ 0.74 & - & 0.74 & 1.74 & 1.74 \\ 3.32 & 0.74 & - & 0.74 & 0.74 \\ 0.74 & 1.74 & 0.74 & - & 1.74 \\ 0.74 & 1.74 & 0.74 & 1.74 & - \end{bmatrix}$$

Diagonal values are usually defined by convention to be slightly larger than the maximum element, to preserve the positivity axiom of a metric.

3 | Scaling: The Basic Non-Metric Distance Model

The journey of a thousand miles begins with a single step.

LAO TZE

3.1 Ordinal Rescaling: Introduction

In using MDS, the user must pay attention to three things:

the data, which give empirical information on how the objects or stimuli relate to each other;

the model, which provides a set of assumptions in terms of which the data will be interpreted; and

the transformation, which is the rescaling which may legitimately be performed on the data to bring them into closer conformity to the model. This is usually referred to as the 'level of measurement' of the data.

The data

In the basic MDS model (frequently called 'smallest space analysis' or 'non-metric distance scaling') the data take the form of a square, symmetric 2-way table, whose entries indicate how similar or how dissimilar any two points are. (To avoid unnecessary repetition we shall assume that the data are dissimilarities, unless otherwise indicated). By convention, the entry in the ith row and jth column of the table is denoted δ_{ij}, and gives the value of the dissimilarity measure between object i and object j. Because the data are symmetric ($\delta_{ij} = \delta_{ji}$) and each object is considered to be identical to itself, the diagonal entries are ignored, and only one half of the matrix, usually the triangle below the diagonal, is presented (see Table 1.1 as an example).

The model

The model used in basic MDS is the simple Euclidean distance model described in section 2.1. In terms of this model, the data δ_{ij} will be interpreted as being 'distance-like'; not as actual distances, but as approximate or distorted estimates of distance. The aim of the MDS analysis is to turn such data into a set of genuine Euclidean distances. The solution (also called the 'final configuration') consists of an arrangement of points in a small number of dimensions, located so that the distance between the points matches the dissimilarities between the objects as closely as possible.

Column:	(1)	(2)	(3)	(4)	(5)	(6)
			DATA		DISTANCES	
	No.	Pair	Data	Rank	Real*	(Est.) Scaled**
		(i, j)	$\delta(i, j)$	$\rho(i, j)$	$d(i, j)$	$d(i, j)$
	1	(8, 2)	0.932	1	5.830	2.411
	2	(8, 1)	0.901	2	5.656	2.270
	3	(6, 1)	0.899	3	5.385	2.206
	4	(7, 6)	0.833	4	5.099	2.206
	5	(7, 3)	0.752 ⎱ 5 =		5.000	2.145
	6	(6, 2)	0.752 ⎰		5.000	2.118
	7	(7, 2)	0.730 ⎱ 7 =		4.123	1.827
	8	(8, 7)	0.730 ⎰		4.123	1.739
	9	(3, 1)	0.712 ⎱ 9 =		4.000	1.632
	10	(8, 3)	0.712 ⎰		4.000	1.558
	11	(8, 4)	0.634	11	3.605	1.557
	12	(6, 4)	0.541 ⎤		3.162	1.429
	13	(5, 2)	0.541 ⎬ 12 =		3.162	1.377
	14	(3, 2)	0.541 ⎦		3.162	1.367
	15	(6, 5)	0.530 ⎱ 15 =		3.000	1.320
	16	(7, 1)	0.530 ⎰		3.000	1.308
	17	(7, 4)	0.521 ⎤		2.828	1.264
	18	(5, 3)	0.521 ⎬ 17 =		2.828	1.221
	19	(8, 5)	0.521 ⎮		2.828	1.148
	20	(5, 1)	0.521 ⎦		2.828	1.130
	21	(4, 3)	0.364 ⎤		2.236	0.972
	22	(7, 5)	0.364 ⎮		2.236	0.941
	23	(8, 6)	0.364 ⎬ 21 =		2.236	0.895
	24	(6, 3)	0.364 ⎮		2.236	0.882
	25	(4, 2)	0.364 ⎮		2.236	0.866
	26	(4, 1)	0.364 ⎦		2.236	0.779
	27	(2, 1)	0.211	27	1.414	0.586
	28	(5, 4)	0.007	28	1.000	0.532

*Strong monotonicity, secondary approach to ties
**Weak monotonicity, primary approach to ties

Table 3.1   *Ordinal rescaling of the data from Table 1.1*

The transformation
The ordinal or monotonic* transformation used in non-metric MDS assumes that only the rank order of the entries in the data matrix contains significant information. Consequently the distances of the solution should, as far as possible, be in the same rank order as the original data. For this reason, non-metric MDS is sometimes referred to as 'ordinal rescaling analysis' (Sibson 1972).

The purpose of the basic non-metric MDS procedure, then, is to find a configuration of points whose distances reflect as closely as possible the rank order of the data. This

*A *monotone* (or monotonic) *increasing* quantity is one which never decreases. (a *monotone decreasing* quantity is one which never increases). Hence, a monotonic transformation of data preserves their order, and in this text the terms 'monotonic' and 'ordinal' are used interchangeably.

is done by trying to find an ordinal rescaling of the data which transforms them into Euclidean distances.*

3.1.1 Perfect ordinal rescaling

To illustrate ordinal rescaling, let us return to the data originally presented in Table 1.1. First, the 28 entries of the data matrix are sorted into order. Column 2 of Table 3.1 gives the (column, row) location of each entry in the matrix, and column 3 gives the actual dissimilarity value (the data). Thus the highest dissimilarity, between object 8 (receiving) and object 2 (rape), has a value of 0.932 and the lowest dissimilarity, between object 5 (libel) and object 4 (perjury), has a value of 0.007. The rank number of each data entry is given in column 4. Note that there is a goodly number of tied data values, including six with the same value of 0.364. In all, only 14 distinct values appear in the data.

In column 5, the original data have been ordinally rescaled into a set of Euclidean distances which correspond to the two-dimensional configuration presented in Figure 3.1. (How the ordinal rescaling was obtained and how the configuration was produced need not concern us at this point. It is only important to see that a configuration has been obtained whose distances are a perfect rescaling of the original data).

The Shepard diagram

It is always instructive to look at the shape of the ordinal transformation function. This is done by producing a 'Shepard diagram' (named after Shepard's seminal paper of 1962), where the data dissimilarities and the distances of the solution are first plotted against each other, and then the ordinal transformation is depicted by joining the points in an upward direction, as is done in Figure 3.2. Since we are dealing with interpoint distances rather than co-ordinates, there will be $p(p - 1)/2$ values contained in the Shepard diagram; with eight points, as here, there are 28 such values. For instance, the bottom left hand point is the one corresponding to (5, 4), whose dissimilarity value is 0.007, and the corresponding distance value is 1.000. The next point up corresponds to (2, 1), with $\delta_{21} = 0.211$ and $d_{21} = 1.414$, and there then follows a point representing the six entries whose data value is 0.364 and whose distance value is 2.236. In all, there are clearly 14 distinct data and distance values. When the points are joined, it can be seen that the transformation function between the data dissimilarities and the solution distances is perfectly monotone, i.e. always moves upwards and to the right. (If the data were similarities, the direction of transformation function would be downward and to the right).

Strong and weak monotonicity

This particular rescaling function illustrates a *strong* (or *strict*) monotonic relationship between the data and the distances, defined in the following way:

> **Strong monotonicity:** Whenever $\delta_{ij} < \delta_{kl}$ then $d_{ij} < d_{kl}$

*The exposition of the basic MDS model in the following sections is primarily based upon the Shepard (1962) and Kruskal (1964a, b) procedure for non-metric MDS, supplemented by the work of Guttman (1968) and Lingoes and Roskam (1973).

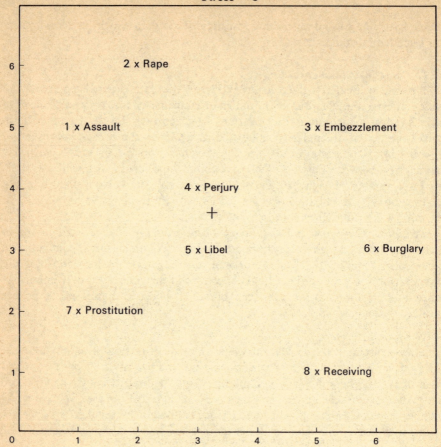

Figure 3.1 *Two-dimensional configuration generating distances of Table 3.1*

That is, if one datum is less than another then the corresponding distances must be in the same order.

A less restrictive requirement often encountered in scaling is that a *weak* monotonic relationship holds between data and distances. Weak monotonicity only requires that no *inversions* in order should occur between the data and distance. That is, that if $\delta_{ij} < \delta_{kl}$ then it should never be the case that $d_{ij} > d_{kl}$. However, in the case of weak monotonicity note that d_{ij} may equal d_{kl}, even when $\delta_{ij} < \delta_{kl}$.

> **Weak monotonicity:** Whenever $\delta_{ij} < \delta_{kl}$ then $d_{ij} \leqslant d_{kl}$

In this case, the transformation function moves upward (even vertically upward) but it may never move downwards. Figure 3.6 gives an example of a weak monotonic transformation of the same data.

3.1.2 *An illustrative example: Scottish mileages*
People often need convincing that it is really possible to derive distance information purely from the rank order of pairwise dissimilarities. A further, less artificial, example should persuade doubters. (In addition, the following example

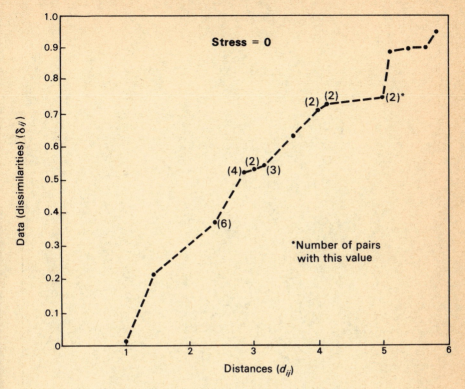

Figure 3.2   *Strong monotone rescaling of data*

involves a monotonic transformation with a regular or smooth shape, and serves to introduce some further ideas of 'badness of fit').

Sixteen towns on the mainland of Scotland were chosen, their distances were measured on a map with a ruler, and a small degree of error (inaccuracy) was added to the distances. The 120 distances were then reduced to rank order, and the rank numbers used as data. These are presented in the bottom left hand corner of Figure 3.3. (There are 92 distinct values, so the number of tied data values is much less than in the previous example).

These data were submitted to MINISSA, the program in the MDS(X) series implementing the basic model, and the configuration presented in Figure 3.3 was produced. (The map outline is drawn in freehand, and the dimensions were rotated counterclockwise through 90° to give the northern orientation to the configuration.) Obviously, it is an excellent recovery of the original configuration of 16 towns. The corresponding Shepard diagram is presented in Figure 3.4. In this example, the monotonic 'line' has not been drawn in, because the smooth, regular shape of the relationship is clear simply by looking at the pattern of the points. The relationship between ranks and distances is linear in the main range (say between 0.5 and 2.00 along the distance axis), but over the entire range the relationship is

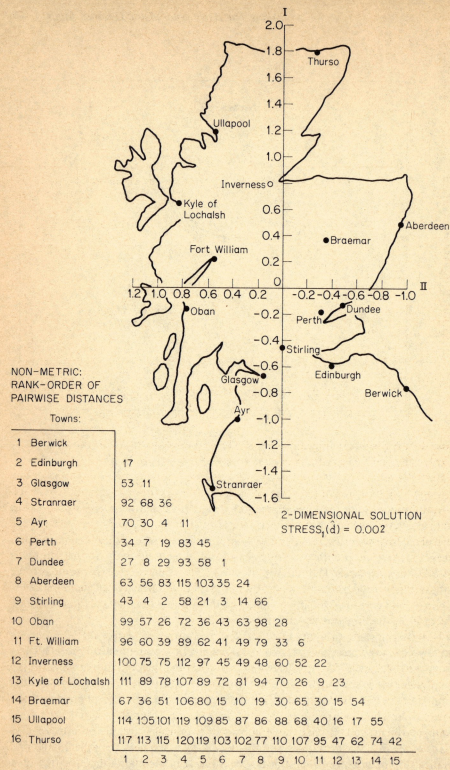

NON-METRIC:
RANK-ORDER OF
PAIRWISE DISTANCES

2-DIMENSIONAL SOLUTION
$STRESS_1(\hat{d}) = 0.002$

Towns:	1	2	3	4	5	6	7	8	9	10	11	12	13	14	15
1 Berwick															
2 Edinburgh	17														
3 Glasgow	53	11													
4 Stranraer	92	68	36												
5 Ayr	70	30	4	11											
6 Perth	34	7	19	83	45										
7 Dundee	27	8	29	93	58	1									
8 Aberdeen	63	56	83	115	103	35	24								
9 Stirling	43	4	2	58	21	3	14	66							
10 Oban	99	57	26	72	36	43	63	98	28						
11 Ft. William	96	60	39	89	62	41	49	79	33	6					
12 Inverness	100	75	75	112	97	45	49	48	60	52	22				
13 Kyle of Lochalsh	111	89	78	107	89	72	81	94	70	26	9	23			
14 Braemar	67	36	51	106	80	15	10	19	30	65	30	15	54		
15 Ullapool	114	105	101	119	109	85	87	86	88	68	40	16	17	55	
16 Thurso	117	113	115	120	119	103	102	77	110	107	95	47	62	74	42

Figure 3.3 *Data (rank of mileages) and solution of Scottish distances.*

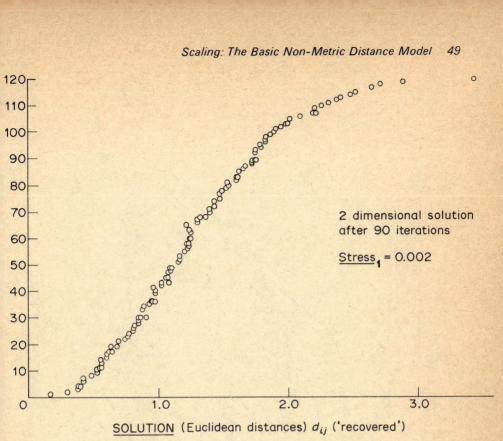

2 dimensional solution
after 90 iterations

Stress$_1$ = 0.002

SOLUTION (Euclidean distances) d_{ij} ('recovered')

Figure 3.4   *Shepard diagram: Scottish distances*

close to an S-shaped (logistic) curve, where the initial smallest values increase slowly, and the final largest values decrease slowly.

But notice that a few rank mileages are not well fit—if a line were to be drawn through all points, it would occasionally have to move *downwards* and to the right to accommodate all the points, contrary to the requirement of monotonicity. True, there are very few such instances, but they are sufficient to show that, even in the case of slightly imperfect error-prone data, it will not always be possible to define a perfect monotonic rescaling. Instead we shall have to talk about fitting or estimating a monotonic 'line', about errors (or departures from a monotonic relationship), and these will be used to define *stress* as an overall measure of fit.

3.2   Monotone Regression : The basic ideas

In using non-metric MDS a perfect ordinal rescaling of the data into distances is usually not possible. What is sought is as good a rescaling as can be achieved. Later, in section 3.4, it will be seen that this involves finding a *series* of configurations in which the interpoint distances come more and more closely into conformity with the data. For the present it simplifies matters to concentrate upon how well one particular configuration (or, strictly, its distances) matches the data. To do so it is crucial to grasp the important, but basically simple, ideas involved in what is termed 'monotonic (or ordinal, or isotonic) regression'. To illustrate the main ideas, we return to the previous example of rescaling the data of Table 3.1, which is illustrated in Figures 3.1 and 3.2.

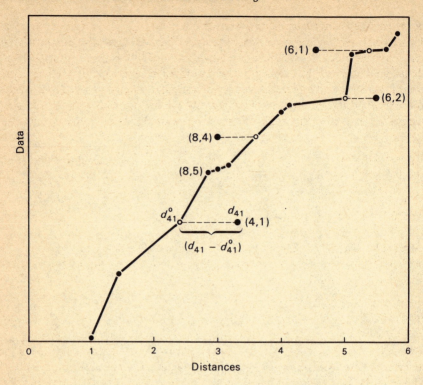

Figure 3.5 *Imperfect monotone rescaling of data*

Suppose points 1, 4 and 6 in the final configuration of this earlier example were moved slightly; the changed configuration would not fit the data so well, and when the Shepard diagram was constructed it would have the characteristic form of Figure 3.5. The distances between the objects (4, 1), (8, 4), (6, 2) and (6, 1) have now changed, and in such a way that the data and the distances can no longer be put into a perfect monotone relationship. For instance, the data indicate that $\delta_{41} < \delta_{85}$, whereas now $d_{41} > d_{85}$. This inversion in ordering means that it is impossible to construct a monotone 'line' through all the (black) points on the Shepard diagram. We are in a quandary: we must either follow the counsel of perfection, and declare that properly speaking an ordinal rescaling is not possible, or recognise the fallibility of the data, and construct as good an ordinal rescaling as possible. In actual fact, it is the second option which is usually followed. But to do so, we shall need to know how to construct 'as good a rescaling as possible', and we shall need a precise definition of what is meant by the 'fallibility' of the data.

This can be accomplished by means of a process of monotone regression. This involves the calculation of a new set of 'distances', often called 'pseudo-distances', (since they are not *actual* distances corresponding to any real configuration nor, indeed, need they obey the triangle inequality), or 'fitted distances' or 'disparities'. These three terms are used interchangeably and are denoted: d_{jk}^0. At this point, it is not important to know how they are calculated—that can wait until section 3.5.2. But it is important to know what the properties are and how they serve to measure the extent to which a given configuration fits the data.

The fitting values are ratio-level quantities defined as '*distances*' *which would preserve perfect monotonicity with the data*. Once calculated for each pair of points (j, k), the discrepancy between the actual distance (d_{jk}) and that disparity (d_{jk}^0), which would give a perfect monotonic solution (i.e., $d_{jk} - d_{jk}^0$), serves as a basis for measuring how far the distances of the configuration depart from those 'pseudo-distances' necessary to keep perfect order with the data. Note that these differences between the distances and the fitting values are calculated along the *distance* axis. The reason for this is that we have committed ourselves to regarding the data as ordinal and any arithmetic operations involving them (in this case, subtraction) are illegitimate. In Figure 3.5 the disparities are denoted by white circles, in the cases where they differ from the actual distances.

3.2.1 Two types of fitting value

In non-metric MDS, two types of fitting quantities (disparities) are frequently used—one of which measures the deviation of the distances of the configuration from *weak* monotonicity, and the other which measures deviation from *strong* monotonicity. Both of these types are used in many MDS(X) programs, and it is important to see how they differ, and that strong monotonicity is usually bound to produce worse fit than weak monotonicity.

(i) *Weak monotonicity* (Kruskal 1964a)

In this case the pseudo-distances which are fitted in the monotone regression are denoted $\hat{d}_{jk}$ ('*d*-hat'), and are required to be *weakly* monotone with the data:

> **Weak mon:** Whenever $\delta_{ij} < \delta_{kl}$ then $\hat{d}_{ij} \leqslant \hat{d}_{kl}$

That is, weak monotonicity allows unequal data to be fitted by equal disparities. When this happens it shows up on a Shepard diagram in the form of vertical lines in the monotone transformation function. In addition the disparity values, $\hat{d}_{jk}$ have the useful property of being as close as possible to the corresponding distances. This means that, over all the points of the configuration, the sum of the squared differences between the distances and the corresponding disparities is as small as possible, i.e.

$$\sum_{\substack{\text{all pairs} \\ (j, k)}} (d_{jk} - \hat{d}_{jk})^2 = \text{minimum}$$

In brief, Kruskal's fitting quantities, also referred to as BFMF (best-fitting monotone function estimates), are required to be both weakly monotone with the data and a least-squares fit to the actual distances.

(ii) *Strong monotonicity* (Guttman 1968)

In this case the pseudo-distances fitted in monotone regression are denoted d_{jk}^* ('*d*-star'), and are required to be *strongly* monotone with the data. They are often referred to as 'rank images' estimates:

> **Strong mon:** Whenever $\delta_{ij} < \delta_{kl}$ then $d_{ij}^* < d_{kl}^*$

It should be noted that strong monotonicity does *not* allow unequal data to be fitted by equal disparities. In this case no vertical segments will appear on the Shepard diagram. Consequently, if the criterion of strong monotonicity is chosen, more discrepancies will occur between the data and the distances of the solution than if weak monotonicity is chosen. Moreover, it is not required that the $d*$ values be as close as possible to the actual distances, so the difference $(d_{jk} - d_{jk}^*)$ will usually be larger than $(d_{jk} - \hat{d}_{jk})$.

As a result of these differences in definition, any overall measure of badness-of-fit between a particular configuration and a set of data is bound to be higher (i.e. worse-fit) when measured in terms of departure from Guttman's strong monotonicity requirement than when measured from Kruskal's weak monotonicity requirement.

In using MDS programs it is important to pay attention to which form of monotonic regression is being used. A full technical comparison of the differences and similarities is contained in Lingoes and Roskam (1973) and in Young (1973).†

3.2.2   Ties in the data

In a set of research data, the values will not normally be distinct; at least some values will be the same. The question arises: should equal dissimilarities be fit by equal disparities?

Two main answers have been given to the question, referred to as the 'primary' and the 'secondary' approach to ties:

Primary approach

> **Primary approach to ties:** If $\delta_{ij} = \delta_{kl}$ then d_{ij}^0 may, or may not, equal d_{kl}^0

This indulgent approach treats ties as indeterminate and allows fitting values either to preserve the equality or replace it by an inequality. In fact, the tie will be broken if in so doing the goodness of fit is improved. In a Shepard diagram, the primary approach to ties shows up characteristically in the form of horizontal straight lines in the monotone function (since identical dissimilarity values are allowed to be represented by different distance values). Figure 3.6 presents a perfect weak monotonic rescaling of the data of Table 3.1, *using the primary approach to ties.* Compare this with the strong monotonic rescaling in Figure 3.2. The configuration recovered by the two scalings is however, virtually identical.

Secondary approach

> **Secondary approach to ties:** Whenever $\delta_{ij} = \delta_{kl}$ then $d_{ij}^0 = d_{kl}^0$

On the other hand, in the secondary approach, ties in the data are required to be retained in the fitting values. Consequently, if the actual distances do not preserve

---

†Young shows that Kruskal's weak monotone and Guttman's rank image transformations mark the extremes of a continuum of a bounded, one-parameter, family of possible monotonic transformations. Although other possible variants have some desirable properties, they are not used extensively in MDS programs and they all seem able to recover metric information with about equal proficiency.

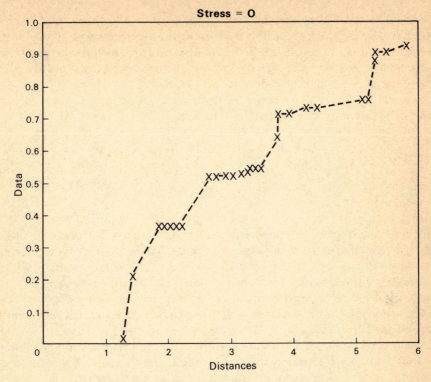

Figure 3.6 *Weak monotone rescaling of data*

every equality in the data, each infraction will be counted as a deviation from monotonicity. In effect, in the case of the secondary approach tied data are treated as being genuinely equivalent. (Note from Table 3.1 that Figure 3.2 is a perfect strong rescaling, using the secondary approach to ties.)

In general, the primary approach to ties should be used in preference to the secondary approach, especially if there is a fairly large number of distinct values in the data. Kendall (1971b, p. 313 et seq.) shows that adoption of the secondary approach can badly misrepresent the structure present. However, if there is only a small number of distinct dissimilarity values (as, for instance, when the data are ratings of similarity from a scale containing a very limited number of ordered categories) then allowing the program the additional indulgence of fitting equivalent category values by disparities in any order may destroy virtually all information.

Generally, MDS programs use the primary approach to tied data in obtaining a solution and, in the MDS(X) series, MINISSA(N), among others, offers the user the choice of primary or secondary approach.

The decision as to what values count as the same is far from trivial. Often data input will consist of numerical association coefficients, such as correlations, which have been calculated to several decimal places of accuracy. So long as two values differ—even if only in the final place—a non-metric program will treat them as distinct and attempt to find corresponding distance values which are also distinct. Such spurious exactness can be avoided by including coefficient values only up to the desired level of exactness (e.g. by rewriting an INPUT FORMAT of: 10F8.5 as, say, 10(F5.2, 3X) in order to keep only two significant decimal places). Another alternative is to employ the parameter EPSILON in MINISSA(N) or the parameters TIEUP and/or TIEDOWN in SSA(M); in either case values will be treated as identical if they differ by less than a specified amount (see also 6.1.1).

3.2.3 *Preservation of order information*

Taken together, the monotonicity criterion (weak *vs* strong) and the approach to ties (primary *vs* secondary) produce somewhat different effects on the preservation of ordinal information (order inequalities and equal values) in the data, and the monotonic function has a slightly different form in each of the four cases. These alternatives are presented as a typology in Table 3.2.

Type I: weak monotonicity
This is the most commonly employed option, and the most indulgent one, since it allows maximum flexibility in rescaling the data ordinally. At best it can recover structure which is obscured by a good deal of error (cf. Kendall 1971b); at worst it can destroy virtually all significant information if it ties data which ought not to be equal and unties data which ought to be ordered.

   Characteristically, the monotone regression function in this case is very 'steppy', with a number of both vertical segments (due to weak monotonicity allowing different data values to have the identical fitting value), and horizontal segments (due to the primary approach allowing identical data values to be fit by different differences). See Figure 3.6 for an example of this, although the only effect of weak monotonicity occurs in the ·single small vertical segment in the top right-hand corner. On the other hand, there are a number of obvious instances of the effect of primary ties (horizontal segments). In general, type I monotone regression functions are 'upward non-decreasing', and often include a number of right-angle steps.

Type II: semi-weak monotonicity
No horizontal segments appear on the function, since they are excluded by the secondary approach to ties. But vertical segments will usually occur. It is a combination which should be used when the user ascribes greater importance to tied information in the data than to the order information.

Type III: semi-strong monotonicity
In this case, no vertical segments occur on the function, since strong monotonicity precludes them. But the function will normally contain horizontal 'plateaux', indicating the operation of the primary tying option. This combination is used surprisingly infrequently, given that it seeks to preserve the significant order information, but gives the freedom to treat ties as indeterminate.

Type IV: strong monotonicity
This is the most restrictive option, where the function is strictly increasing, but can never be a step function. It should be used when the data are believed (or known) to be virtually error-free. An example appears in Figures 3.2 and 3.5. It can also serve as a salutary reminder of what a rigorous and uncompromising interpretation of ordinal measurement actually involves when applied to real data,

TYING OPTION CHOSEN

MONOTONICITY CRITERION	Primary (Indeterminate)	Secondary (Equivalence)
Weak *(Kruskal's $\hat{d}$)*	I WEAK *May equate unequal data* May untie tied data	II SEMI-WEAK *May equate unequal data* Preserves ties
Strong *(Guttman's $d*$)*	III SEMI-STRONG *Preserves strict* *inequalities* ($>$, $<$) May untie tied data	IV STRONG *Preserves strict* *inequalities* ($>$, $<$) Preserves ties

Table 3.2 *Preservation of ordinal information in monotone regression*

Note This table and terminology are based upon Roskam (1969, pp. 9–11) and is reproduced with permission.

since every infraction will count as evidence against the hypothesis that the data can be perfectly rescaled ordinally into Euclidean distances.

Note that the question of how to calculate the fitting values has so far been ignored, but is taken up in 3.5.2. Monotonic regression may also be performed on quantities other than distances: it provides a general procedure for comparing the ordinal rescaling of data into a corresponding set of quantities defined by *any* sort of model (e.g., factor or scalar-product models, additive models, etc.; see Chapter 5).

3.3 Goodness/Badness of Fit: Stress and Alienation

We have seen that the difference between a particular distance and its corresponding 'pseudo-distance' ($d_{jk} - d_{jk}^0$) serves as an index of how badly the distance between j and k in the solution configuration departs from the value required to preserve an ordinal relation with the data. If there is no inversion in the required ordering then the difference will be zero. Alternatively, the difference can be looked on as the residual from monotone regression, i.e. an index of the difference between the solution distance and (an ordinal rescaling of) the data.

A simple *overall* measure of how the distances in a configuration ordinally fit the data can be constructed by squaring the differences between the actual distances in the configuration and the 'distances' fitted by monotone regression, and then sum them. MDS almost universally adopts the habit of using a *badness*-of-fit measure—the higher the index, the worse the fit—to assess the fit between the solution and the data. This basic index, called variously raw stress (Kruskal), raw phi (Guttman, Lingoes, Roskam), or stressform 0 has the same form as the 'residual sum of squares' in other types of regression, except that in this case it measures the residuals from monotonic regression. We shall refer to it normally as raw stress.

$$S_0 = \begin{cases} \text{Stressform 0} \\ \text{Raw stress} \\ \text{Raw phi} \end{cases} = \sum_{\substack{\text{all pairs} \\ (j,k)}} (d_{jk} - d_{jk}^0)^2$$

By convention, if the fitting quantities are Kruskal's $\hat{d}_{jk}$, then S_0 is referred to as raw stress, and if Guttman's d_{jk}^* are used, then it is called raw phi. In any event, for the same configuration, raw phi based on rank images will normally be higher than raw stress based on Kruskal's BFMF quantities, because of the strong monotonicity requirement. That is:

$$S_0(d^*) \geqslant S_0(\hat{d})$$

Raw stress is unfortunately a very unsatisfactory measure of fit for MDS solutions. The reason is that configurations which are identical in all but size will have different values of raw stress. But it is not the actual numerical distances (or co-ordinates) of an MDS configuration which are important or significant, but only the *relative* distances. For instance, doubling or halving the scale of the configuration is usually considered simply an irrelevance. We are only concerned with obtaining a configuration of points which is unique up to the uniform stretching or shrinking of the axes (or distances) by *any* constant, which is simply another way of saying that distances are at the ratio level of measurement. But unfortunately, if a configuration is shrunk uniformly by a constant, k, then the raw stress value shrinks by a value of k^2. That being so, if raw stress is used as an index of fit, it will always be possible to get a better fit simply by scaling down the size of the configuration! This is obviously an undesirable state of affairs, but the remedy is simple. To prevent it happening, raw stress can be divided by a factor which takes the size of the configuration into account, which has the effect of giving the same stress value to all configurations which differ only in size. A number of such 'normalising' or 'scale' factors have been proposed (see Kruskal and Carroll 1969, Roskam 1975, Lingoes and Roskam 1973). One family (the 'stress' indexes) stems largely from the Bell Laboratories group, and another family from the Guttman-Lingoes-Roskam group. Since programs from both sources are included in the MDS(X) series, the interrelations of these various measures are discussed in Appendix A3.1. For expository purposes, it will be sufficient to discuss the two most commonly used versions of normalised stress, each of which is widely used.

3.3.1 Normalised forms of stress
By normalising raw stress, it is possible to compare configurations by making stress independent of the size or scale of the configuration, and norming its value between 0 (perfect fit) and 1 (worst possible fit). The two most commonly used normalising factors are:

NF 1:

$$\sum_{(j,k)} d_{jk}^2 \quad \text{(the sum of the squared distances)}$$

This removes dependence on the scale of the distances. Its value will equal p^2, where p is the number of stimuli, if the configuration is centred and standardised (which means that the sum of co-ordinates of each dimension is zero and the sum of the squared co-ordinates is p) as is normally the case in MDS solutions.

NF 2:

$$\sum_{(j,k)} (d_{jk} - \bar{d})^2 \quad \begin{array}{l}\text{(the sum of the squared differences between} \\ \text{the distances and their average, } \bar{d})\end{array}$$

This factor represents the variation of the distances about their mean and should always be used when data are conditional, such as ratings or rank orders (see 5.6.2), since it helps prevent a situation where a subject's rank values are fitted by the same disparity value (see Kruskal 1965).

Basically, normalised stress measures take the form:

RAW STRESS/NORMALISING FACTOR

but usually the square root of this ratio is taken, which has the effect of deflating the size of the index and making it sensitive to relatively small improvements when a configuration is coming close to being a perfect fit to the data (cf. Roskam 1968, pp. 34–5).

$$S_1 = \begin{cases} \text{Stressform 1} \\ Stress_1 \\ \sqrt{(2 \times \text{phi})} \end{cases} = \sqrt{(\text{Raw Stress/NF 1})}$$

$$= \sqrt{\left\{ \sum \left(d_{jk} - d_{jk}^0 \right)^2 \Big/ \sum d_{jk}^2 \right\}}$$

3.3.1.1 Properties of stress

Several important properties of stress$_1$ become more obvious if we consider its squared values, S_1^2:

(i) when S_1 (and hence S_1^2) is zero, then there is perfect ordinal fit, and all fitted 'distances' will equal the actual distances,

$$d_{jk}^0 = d_{jk}, \qquad\qquad \text{for all } (j, k)$$

(ii) the maximum value of stress$_1$ is more difficult to determine, but S_1^2 can be shown (Lingoes and Roskam 1973, p. 12) to reach a maximum of $(1 - 2/p)$, which implies that S_1 approaches 1 as an upper limit as p, the number of stimuli, gets larger (as $p = 8, 16, 32, 64$; S_1 (max) $= 0.87, 0.94, 0.97, 0.98$).

(iii) S_1^2 can be re-expressed as

$$S_1^2 = 1 - \left(\sum d_{jk}^0 \Big/ \sum d_{jk} \right)$$

which is equivalent to the proportion of residual variance from monotone regression.

$$S_2 = \begin{cases} \text{Stressform 2} \\ Stress_2 \end{cases} = \sqrt{(\text{Raw Stress})/\text{NF 2}}$$

$$= \sqrt{\left\{ \sum \left(d_{jk} - d_{jk}^0 \right)^2 \Big/ \sum \left(d_{jk} - \bar{d} \right)^2 \right\}}$$

Stress$_2$ can be interpreted as being the variation between the distances and the disparities as a fraction of the variation of distance round their mean. It has the same properties as stress$_1$ when zero, but its maximum value is difficult to determine. In general stress$_2$ will be larger than stress$_1$, often twice as large, for the same configuration. Although stress$_2$ should always be chosen in preference to stress$_1$, for conditional data it makes little practical difference which is used to monitor an MDS solution in the case of the basic model.

So far we have dealt with a situation in which we have a set of data and a configuration. We have sought to measure how closely the distances between the points in that configuration are to a monotonic rescaling of the data, and the measure stress, which performs this task, has been described. We now go on to see how stress can be used to indicate not only how well a particular configuration captures the information in the data but also how an imperfectly fitting configuration can be improved to fit the data.

3.4   Finding the Best Configuration
The next question is: how does non-metric MDS actually work? Given a set of data, how does one find a configuration of points in Euclidean space where the rank-order of the distances best matches the rank order of the data?

3.4.1   *Data as constraints on the solution*
In principle, it ought to be possible to find a solution analytically. The rank order of the data imposes a set of constraints on where the points can be positioned in a configuration if it is going to conform to the data. As the number of points increases, information on the rank order of the dissimilarities begins to constrain the location of the points in the configuration so much that the distances to all intents and

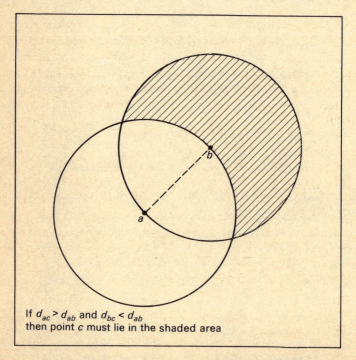

If $d_{ac} > d_{ab}$ and $d_{bc} < d_{ab}$
then point c must lie in the shaded area

Figure 3.7   *Constraints on positioning a point*

purposes become fixed. (This point was also made by Abelson and Tukey (1959) and by Shepard (1962b, p. 238 et seq.)).

Actually, the position of the point never becomes 'fixed', but rather becomes constrained within a smaller and smaller region of the space. Consider a two-dimensional plane. Suppose there are two points a and b, then a third point, c, whose distance from a is greater than that between a and b (i.e. $d_{ac} > d_{ab}$), must lie outside of the circle whose centre is a and whose radius is d_{ab} (see Figure 3.7). If we also know that d_{bc} is less than d_{ab} then, in addition, c must lie *within* the circle whose centre is b and whose radius, again, is d_{ab}. Thus c must lie in the shaded area in Figure 3.7, but may be located anywhere within it.

It should be clear that as the number of points increases and the number of such inequality constraints also increases, then the area within which a point may be positioned in accordance with these constraints becomes smaller and smaller (and indeed may not exist). This region within which a point must lie, bounded by the inequality constraints implied by the data, is known as an *isotonic region*, since any point within the region equally satisfies the constraints. As Shepard puts it:

> Actually, though, if non-metric constraints are imposed in sufficient number, they begin to act like metric constraints. In the case of a purely ordinal scale, the non-metric constraints are relatively few and, consequently, the points on the scale can be moved about quite extensively without violating the inequalities (i.e. without interchanging any two points). As these same points are forced to satisfy more and more inequalities on the interpoint distances as well, however, the spacing tightens up until any but very small perturbations of the points will usually violate one or more of the inequalities.*
>
> (Shepard 1966, p. 288)

The imposition of more and more constraints by increasing the number of points hence means that the solution becomes more fixed or determinate. What is really happening is that the rank order of the $p(p-1)/2$ dissimilarity coefficients in the data is being 'distilled' or concentrated into the very much smaller $(p \times r)$ co-ordinates needed to define the solution configuration, and the number of data increases much faster than the number of co-ordinates, so long as the dimensionality is small.

3.4.2 The solution as an iterative process

How is the solution configuration obtained? It is done by a process which is surprisingly simple in form, though sometimes technically complex in detail. The user begins by choosing the number of dimensions r in which she wants a solution to be obtained. Roughly speaking, there should be at least twice as many data as parameters needed to specify the configuration, i.e. $\{p(p-1)/2\} > 2(p \times r)$, so the choice of dimensionality of the solution should be made with this in mind. Other factors are also relevant, and are considered below in section 3.7.

The basic outline of the process used to obtain a solution is as follows:

---

*Suppes and Winet (1955) had already established that for the *unidimensional* case a *complete* ordering of all pairwise distances on a closed interval establishes the representation of these distances up to a multiplicative constant—i.e. a complete 'ordered metric scale' is equivalent in the limit to a *ratio* scale. (Strictly, this only applies for the limiting case of an infinite number of points). This also confirms the similar conjecture by Abelson and Tukey (1959).

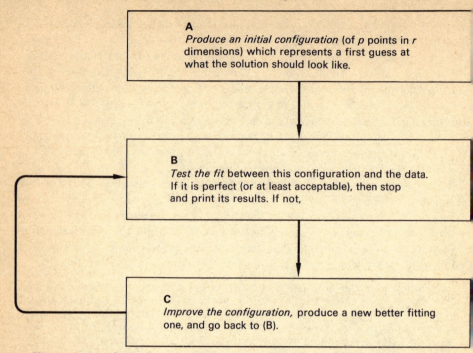

A
Produce an initial configuration (of *p* points in *r* dimensions) which represents a first guess at what the solution should look like.

B
Test the fit between this configuration and the data. If it is perfect (or at least acceptable), then stop and print its results. If not,

C
Improve the configuration, produce a new better fitting one, and go back to (B).

The process of moving round the cycle from C to B, producing somewhat improved configurations each time, is termed an 'iterative procedure', each cycle being an iteration. Before the advent of electronic computers, such procedures were simply not feasible—it is not unusual for there to be 50–100 iterations before a satisfactory solution is obtained.

All non-metric MDS programs follow this same basic procedure, but there are considerable differences in the details of the process, many of them now of purely historical interest. Very similar procedures had also been developed independently in Japan by Hayashi (1968) termed 'quantification scaling', and in France by Benzécri (1964) who had produced 'l'analyse des correspondances'.

3.5   General Outline of MINISSA
The next step is to outline the general iterative procedure in a little more detail, but in a non-technical manner. The overall flow is illustrated in Figure 3.8.

Let us begin by expanding on the 3-step process to include the major steps followed in all the basic programs.

3.5.1   The initial configuration
(i)*   **Create an initial configuration** which will provide a good estimate of the final solution. This will reduce the number of iterations, and lower the probability of finishing with a sub-optimal solution (or 'local minimum', see section 3.5.4). In the case of MINISSA, this is done by keeping the *r* most important principal components of a matrix which is based upon the rank-order of the dissimilarities (see Appendix A3.2 for details).

---

*The roman numerals follow the steps of the sequence summarised in Figure 3.8.

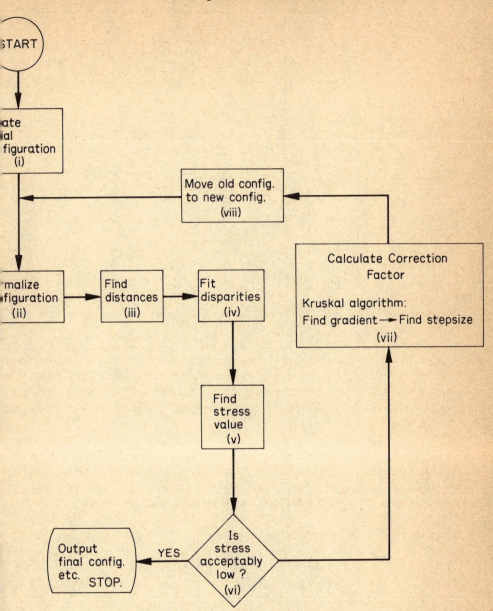

Figure 3.8 *Summary of the iterative process*

(ii) **Normalise this configuration** so that the origin is at the centroid (centre of gravity) of the stimulus point locations, and the configuration has constant dispersion (i.e. the sum of squares of the co-ordinates equals the number of points). This step is not always necessary, but is relevant and important if raw stress or phi is being used to measure badness of fit.

3.5.2 *Comparing the current configuration with the data*
The aim is to produce a configuration whose distances match the rank order of the

I Original data

Stimulus no:	1	2	3	4	5
1	—				
2	3	—			
3	6	4	—		
4	10	5	7	—	
5	2	8	9	1	—

II Sorted into order

	(5,4)	(5,1)	(2,1)	(3,2)	(4,2)	(3,1)	(4,3)	(5,2)	(5,3)	(4,1)		Table Row
(column, row) indices											← (j,k)	1
Data (δ_{jk})	1	2	3	4	5	6	7	8	9	10	← δ_{jk}	2
Distances in *current configuration* (d_{jk})	3	6	3	5	8	10	13	11	9	15	← d_{jk}	3

III Calculation of disparities

(a) MONOTONE REGRESSION

step	(5,4)	(5,1)	(2,1)	(3,2)	(4,2)	(3,1)	(4,3)	(5,2)	(5,3)	(4,1)		Table Row
1	3											4
2		$4\frac{1}{2}$	$4\frac{1}{2}$									5
3				5	8	10						6
4							11	11				7
5							11	11	11	15		8
Disparities ($\hat{d}_{jk}$)	3	$(4\frac{1}{2})$	$(4\frac{1}{2})$	5	8	10	(11)	11	11	15	$\hat{d}_{jk}$	9

(b) RANK IMAGES (Permutation of d_{jk})

	(5,4)	(5,1)	(2,1)	(3,2)	(4,2)	(3,1)	(4,3)	(5,2)	(5,3)	(4,1)		Table Row
Disparities (d^{*}_{jk})	3	3	5	6	8	9	10	11	13	15	← d^{*}_{jk}	10

IV Squared differences

	(5,4)	(5,1)	(2,1)	(3,2)	(4,2)	(3,1)	(4,3)	(5,2)	(5,3)	(4,1)		Table Row
(a) $(d_{jk} - \hat{d}_{jk})^2$	0	9/4	9/4	0	0	0	4	0	4	0	(Sum): 12.5	11
(b) $(d_{jk} - d^{*}_{jk})^2$	0	9	4	1	0	1	9	0	16	0	(Sum): 40.0	12

V Badness of Fit Measures

			Table Row
Raw Stress ($\hat{d}$)	$= \sum (d_{jk} - \hat{d}_{jk})^2$	$= 12.5$	13
Raw Phi/Stress (d^{*})	$= \sum (d_{jk} - d^{*}_{jk})^2$	$= 40.0$	14
Stress$_1$ ($\hat{d}$)	$= \sqrt{(\text{Raw Stress}/\text{NF }1)}$	$= 0.1221$	15
Stress$_1$ (d^{*})	$= \sqrt{(\text{Raw Stress}/\text{NF }1)}$	$= 0.2184$	16
Stress$_2$ ($\hat{d}$)	$= \sqrt{(\text{Raw Stress}/\text{NF }2)}$	$= 0.2886$	17
Stress$_2$ (d^{*})	$= \sqrt{(\text{Raw Stress}/\text{NF }2)}$	$= 0.5162$	18

NF 1 $= \sum d_{jk}^2 = 839$; NF 2 $= \sum (d_{jk} - \bar{d})^2 = 150.1$

Table 3.3 *Calculation of disparities*

input data as closely as possible. How well does our current configuration, i.e. at the current iteration, match the data? The answer to this question involves three things:

> calculating the distances between the points in the current configuration;
> comparing these distances with the data (by calculating disparity values); and
> assessing how badly the configuration departs from perfect ordinal fit to the data, i.e. by calculating stress.

Each point is now taken up in turn, in (iii), (iv) and (v) below, and illustrated by reference to a simple example shown in Table 3.3. Let us suppose the data dissimilarities are those given in section I ('Original Data') of Table 3.3. The first step consists of sorting the dissimilarities into ascending order (from the lowest to the highest value), keeping track of the column and row reference of each datum. The sorted data values are given in row 2 of Table 3.3, and the table (column and row) indices of each datum are noted in row 1. It should be stressed that *this information remains fixed throughout the iterative procedure*.

(iii) Find distances. Each time a new configuration is produced, a new set of distances is calculated, according to the usual Euclidean distance formula:

$$d_{jk} = \sqrt{\left\{ \sum \left(x_{ja} - x_{ka} \right)^2 \right\}}$$

These current distance values are then slotted into the same position as their corresponding data value, as in row 3. For instance, the distance between points 5 and 1, namely 6, is inserted into the second position which corresponds to the data dissimilarity between stimuli 5 and 1.†

(iv) Fit disparities. Now we are in a position to compare the solution (the distances in the current configuration) with the data, which is done by first calculating disparities (d_{jk}^0), which are to be monotonic with the data. As we have seen, it is possible to calculate either Kruskal's weak monotonic $\hat{d}_{jk}$ values or Guttman's strong monotonic d_{jk}^* rank image values. MINISSA and its cognates calculate both forms of disparities. Guttman's rank-image method has been found to be useful in avoiding sub-optimal solutions especially at the start of the process, but later in the program a switch is made to Kruskal's weak monotone regression, which provides a smoother, 'finer-honed' approach to obtaining an acceptable solution.

(a) *Monotone Regression* (Kruskal's $\hat{d}$ disparities)
The procedure of finding the Kruskal $\hat{d}$ values uses the *weak* monotonicity criterion that if $\delta_{ij} < \delta_{kl}$ then the corresponding $\hat{d}$ values should be in the same order, but are permitted to be *equal* without the tie counting as an infraction of monotonicity.

In brief, monotone regression consists of working consecutively through the distance values, checking whether they are in the same order as the data. However, when an inversion appears, i.e. where one or more distance values *decreases*, then a 'block' is formed by taking the offending value and the preceding one. These are

---

†The reader will have noticed that the distance values are not genuine distances at all, but numbers chosen to simplify the arithmetic of the example.

averaged until monotonicity is restored between blocks. The sequence is illustrated in rows 4 through 8 of Table 3.3. The sequence consists of repeatedly comparing the distances (row 3) and the data (row 2). It is useful to follow the example in detail.

1   The first two distances are compared to the data, and are found to be in correct (increasing) order, but the third distance *decreases* to the value of 3. At this point, we go back one position (to the second distance), treat the distances corresponding to (5, 1) and (2, 1) as tied, and average them to $4\frac{1}{2}$ (row 5). Now we have three disparities which are weakly monotonic with the data, viz

$$data \quad \ldots 1 \quad 2 \quad 3$$
$$disparities \ldots 3 \quad (4\tfrac{1}{2} \quad 4\tfrac{1}{2}) \quad \text{(cf. row 5)}$$

2   The subsequent four distances increase, but then the eighth (and ninth) values decrease, so once more we backtrack to the last distance which *is* in order— the seventh—and average. The sequence is now

$$data \quad \ldots 6 \quad 7 \quad 8$$
$$disparities \ldots 10 \quad 12 \quad 12$$

So far so good.

3   But then the ninth value decreases, so the backup process continues, still averaging distances until finally a block is formed whose average *does* preserve order, with respect to both the block below it and the one above it. (This is what Kruskal (1964b, pp. 40–1) refers to as a block being 'down-satisfied' and 'up-statisfied' with respect to monotonicity):

$$data \quad \ldots 6 \quad 7 \quad 8 \quad 9 \quad 10$$
$$disparities \ldots 10 \quad (11 \quad 11 \quad 11) \quad 15 \quad \text{(cf. row 9)}$$

The monotone regression procedure of fitting a set of disparities to the data is now complete (row 9). There are seven blocks in all, and a set of disparities has been produced which are now perfectly *weakly* monotonic with the data (compare rows 2 and 9).

(b)   *Rank Images* (Guttman's $d*$ disparities)
In Guttman's approach, we seek a set of quantities which are *strongly* monotonic with the data and once again use the distances in the current configuration as a starting point for calculating disparities. Guttman's procedure consists simply of taking the *current* set of distances (row 3), sorting them into order, and using them as the rank-image fitting quantities, $d*$ (row 10).

Notice that in the example chosen, all data values are distinct. This will rarely happen in practice. The calculation of disparities is modified slightly when ties occur in the data, depending upon whether the primary or secondary approach to ties is chosen by the user. The process of calculation is described in detail in Roskam (1975, p. 12 bis). Note also that in this example there are two equal distances (i.e. those between points 4 and 5 and 1 and 2). These two values become the fitting values corresponding to the pairs (4, 5) and (1, 2), and thus in the Shepard diagram would show up as a vertical segment *even though* strong monotonicity is being sought.

(v) **Find stress value.** We are now in a position to assess how well the current configuration fits the data. This is done by calculating the extent to which the actual distances diverge from the distances-made-to-conform-to-monotonicity, i.e. from the disparities. Since the sum of all the differences $(d_{jk} - d_{jk}^0)$ will be zero, they are first squared.[†] Row 11 presents the squared differences based on $\hat{d}$ (formed by subtracting row 9 from row 3 entries and squaring), and row 12 presents those based on d^* (subtracting row 10 from row 3 and squaring). Whichever fitting procedure is used, the overall picture is very similar: four of the ten data are fit perfectly, and in both cases the main distances contributing to the badness-of-fit are between 5 and 3, 4 and 3, 5 and 1, and 2 and 1. Clearly, points 5, 3 and 1 are especially badly positioned in the current configuration, and will need to be moved to achieve a better-fitting configuration.

The overall badness-of-fit is now calculated by summing the squared differences to form raw stress (based either on $\hat{d}$ (row 13) or on d^* (row 14)). Note that rank image fitting produces higher stress values than monotone regression fitting, because strong monotonicity is the more stringent criterion and is not a least-squares fit to the data.[‡]

To compare configurations, a normalised version of stress is necessary, and these are presented in rows 15 to 18. Stress$_1$ $(\hat{d})$ is the measure most commonly reported in the literature, and is to be preferred at least on the grounds that we have more information on its properties and distribution than for any other measure.

(vi) **Is stress acceptably low?** There are several grounds for terminating the iterative procedure:

(a) if the stress value is zero;
(b) if the stress value is 'acceptably close' to zero;
(c) if the improvement in stress since the last iteration is so little that it does not seem worth continuing.

In the first instance, a perfectly fitting configuration has been obtained, and a perfect rescaling achieved.

In the second instance, a number of guidelines have been given as to what value of stress counts as 'acceptable', (see especially Kruskal (1964b, p. 32) and Roskam (1975, p. 16)). The justification for these values is obscure, and even as rules of thumb they should be treated with considerable caution. A rather different approach to assessing stress values is the so-called 'Monte Carlo' simulation approach, which is discussed in greater detail below in section 3.7.1.

A third criterion for terminating the iterative process is simply that there has been so little improvement in the last few iterations that it is scarcely worth continuing. A good example is provided in Figure 3.9, which charts the progress of just over 200 iterations in reducing stress.

Clearly, there is a dramatic decrease in stress in the first few iterations: from 0.334 at the start to 0.158 at iteration 5, and improvement continues until just after

[†]When using rank images, the differences will not necessarily sum to zero, but the squaring convention is employed to keep comparability with BFMF estimates.

[‡]Raw stress/phi, based on d^* rank image fitting (whimsically christened 'soft squeeze' by Guttman) is used to monitor the first stage of MINISSA (and related programs). Stress$_1$ based on monotone regression (called 'hard squeeze' by Guttman) is used to monitor the second stage of MINISSA.

the 100th iteration. Beyond that point, there is virtually no improvement for the next 100 iterations; it would have saved time and expense to have stopped at the point where the improvement between iterations had become negligible.†

3.5.3 *Improving the configuration*
Having calculated the disparities and the measure of overall fit between the present configuration and the data, we now want to produce a *new* configuration whose distances will approximate more closely to the data. Put slightly differently, we want to move the points in the current configuration in such a way as to decrease the stress value.

As we have seen above, by looking at the differences between the current set of distances and disparities, $(d_{jk} - d_{jk}^0)$, we can tell:

1 *which are the greatest discrepancies* (by the absolute value of the difference); and
2 *which points are involved* (by referring to the row and column references of these values).

The conclusions will depend in part upon which disparity values are used. For example, in Table 3.3,

(a) using monotone regression (row 11), the two greatest differences have the value 2, referring to the pairs (4, 3) and (5, 3);
(b) using rank images (row 12), the greatest difference is (5, 3), with a value of 4, followed by (5, 1) and (4, 3) with a value of 3.

But we can also infer two further pieces of information about each pair of points:

3 *in which direction to move the points to produce a better fit;* and
4 *how far to move them.*

(vii) **Calculate correction factor.** A full and accurate explanation of how this information is obtained involves differential calculus, and will not be discussed here.‡ But it is possible to get a perfectly adequate grasp of the basic process by a little simplification, drawing on Gleason (1969, pp. 6–8) and Spence (1978, p. 192). Let us concentrate upon one point, say point 5 in the present example and its relationship to each other point in the current configuration.

First consider point 5 with respect to point 3. Now imagine a line drawn in the configuration to connect points 5 and 3 (its length will be $d_{53} = 9$). How well does the current positioning of points 5 and 3 correspond to the data? If it were a perfect correspondence, then the difference $(d_{53} - \hat{d}_{53})$ would equal zero. As it is, the difference $(9 - 11) = -2$. This tells us that in order to improve the fit point 5 should be moved *away* from point 3 so as to increase the distance, and that this

---

†In the MDS(X) series, control of the number of iterations is given by the ITERATIONS command (MINISSA, SSAM) which sets the minimum number of iterations before a test is made. From then on, a check is made of each iteration to see whether improvement in stress has been more than a given amount. If not, the iterations are terminated.

‡The method of 'steepest descent' or 'negative gradients', which is used to move the configuration is discussed Kruskal (1964b, pp. 30–9). The methods used in MINISSA are reviewed in the MDS(X) documentation, and a full technical discussion is contained in Lingoes and Roskam pp. 12–33).

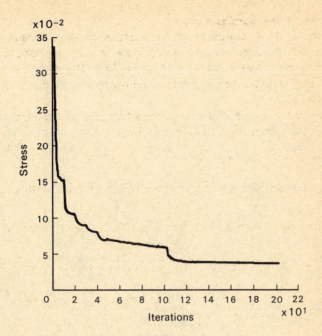

Figure 3.9 *Stress by iteration*

distance should be increased by 2 units if the difference is to be zero. (In general, if $d_{jk} < d_{jk}^0$, then the difference is negative, which indicates that the points should be moved away from each other.)

Now consider point 5 and point 1. The difference $(d_{51} - \hat{d}_{51}) = (6 - 4\frac{1}{2})$ $= +1\frac{1}{2}$. This tells us that point 1 should be moved *towards* point 5 (to reduce the distance, and lower the difference), and should be moved $1\frac{1}{2}$ units if the difference is to be zero. (In general, if $d_{jk} > d_{jk}^0$, then the difference is positive which indicates that the points should be moved towards each other.)

Finally, consider point 5 and its relation to points 2 and 4. In this case, the difference is zero, indicating that the fit is perfect and they should not be moved.

Usually the formula used in MDS programs to improve the position of a point j with respect to another point k takes the form

NEW POSITION of j = OLD POSITION of j + CORRECTION FACTOR

specifically

$$\underset{x_j}{\text{new}} = \underset{x_j}{\text{old}} \qquad + \left(1 - \frac{d_{jk}^0}{d_{jk}}\right)\left(x_j - x_k\right)$$

A little arithmetic reduces the formula to a particularly simple form†:

$$\underset{x_j}{\text{new}} = \underset{x_j}{\text{old}} + (d_{jk} - d_{jk}^0)$$

If there are p points, then for each there will be $(p - 1)$ correction factors, pushing and pulling point p in various directions and with different degrees of force. In this instance, point 5 is being pushed away from point 3 by 2 units, and towards point 1 by $1\frac{1}{2}$ units. The actual move is bound to be a compromise between these various forces, and the greatest discrepancies will obviously tend to dominate the movement of the points.

When all the discrepancies are considered simultaneously, it is necessary to rewrite the correction formula to take into account the forces from *all* the points (represented in the summation), and consider the location of the point p on each dimension, a.

General correction formula:

$$\underset{x_{ja}}{\text{new}} = \underset{x_{ja}}{\text{old}} + \alpha\sum\left(1 - \frac{d_{jk}^0}{d_{jk}}\right)\left(x_{ka} - x_{ja}\right)$$

The only new quantity to appear is α (alpha) the 'step-size' which in Kruskal's version represents the overall amount by which the points are moved. The technicalities are complex,‡ and all the user need know is that longer step-sizes are usually taken in the earlier stages of the process, and when a program is minimising in terms of rank-image disparities, whereas smaller steps are taken towards the end of the process (and when stress is being minimised by reference to Kruskal's weak monotonic disparities). In fact, MINISSA uses a hybrid approach, starting by minimising raw phi/stress in terms of Guttman's Rank Images ('soft squeeze'), and then switching to minimising stress$_1$ in terms of Kruskal's weak monotone disparities ('hard squeeze') later in the process.

---

†The first formula takes the form: $\underset{x_j}{\text{new}} = \underset{x_j}{\text{old}} + (\text{correction factor})$. Now examine the correction factor further. Since we are restricting attention to the line joining x_j and x_k, then $(x_j - x_k)$ is simply d_{jk}. Thus the correction factor reduces to

$$\left(1 - \frac{d_{jk}^0}{d_{jk}}\right)d_{jk}.$$

Putting the term in the brackets over a common denominator, gives

$$\frac{(d_{jk} - d_{jk}^0)}{d_{jk}}d_{jk},$$

and cancelling, the correction factor simplifies to

$$(d_{jk} - d_{jk}^0).$$

Hence, the formula now takes on the simple second form:

$$\underset{x_j}{\text{new}} = \underset{x_j}{\text{old}} + (d_{jk} - d_{jk}^0)$$

‡The basic reference is Kruskal (1964b, p. 121), which is further expanded and discussed in Lingoes and Roskam (1973, pp. 13–16). The Guttman-Lingoes procedure is a somewhat different correction procedure, described in Lingoes and Roskam (1973, pp. 22–9) which provides for different step sizes for each point.

3.5.4 *Local minima and related problems*

Another analogy, used to explain what is happening during the process of moving the configuration to one that fits the data better, is geographical. Suppose our data refer to 20 stimuli and we seek a solution in three-dimensions. We are now asked to think of *all possible* three-dimensional configurations or solutions—good, bad or indifferent—and pay attention to the stress value of each one. Obviously, what we are looking for is that one configuration whose stress is lowest. One way to think about the problem is to simplify matters and imagine the solution space like a 'rolling terrain, with hills and valleys' (Kruskal 1964b, and Kruskal and Wish 1978, pp. 27–8). They continue:

> Each point of the terrain corresponds to an entire configuration (*not* to a point *in* the configuration). Each point of the terrain can be described by three co-ordinates—the altitude, and the two location co-ordinates, North-South and East-West. The locations co-ordinates are analogous to all the co-ordinates of all the points of the configuration. (Of course, a configuration with p points in r dimensional space has $p \times r$ co-ordinates, and $p \times r$ is far greater than two, so the analogy does not convey the full richness and difficulty of the situation). The altitude is analogous to the objective function, that is, the altitude is the stress. (quoted, with slight notational changes, from Kruskal and Wish 1978, p. 27, with permission)

The real problem is that the terrain—the hills of high stress and valleys of low stress—is unknown territory, and we have no means of knowing before the event even what its general features are like, so we are in the position of a blindfolded parachutist dropped from a plane on a dark night (Kruskal and Wish's analogy) or a climber lost in the mist. It is not in fact quite as bad as this—there is a way to locate a sensible starting point, by choosing an initial configuration which gets us fairly close to the point where the stress is lowest. Nonetheless, the imagery is apt. To move from the present position, a configuration of relatively high stress at the earlier stages of the iterative process, we need to know two things: in what direction the ground is sloping downwards; how large a step to take in that direction.

In the first case, we can detect the general direction by calculating the negative gradient, which tells us in what direction to move each point of the current configuration if we want to lower stress; it gives rise to the correction factor formula discussed above in 3.5.3. In the second case, we need to adopt a strategy which avoids the extremes of foolhardiness and over-cautiousness—if we move too far in the general direction of improvement, we might in fact overshoot the actual minimum, and if we move in very small steps we are going to consume an enormous amount of time getting virtually nowhere.

How does the climber know that the minimum (valley floor) has been reached? The answer is, where the gradient is zero; that is, where stress increases in every direction. (This state of affairs is actually tested for at each iteration, and provides yet another criterion for terminating the iterative process.) But unfortunately there is no guarantee, even if a valley floor is reached, that it is actually the lowest point on the terrain—it may simply be a local dip. The situation is illustrated in Figure 3.10. Indeed, there is no sure way of knowing whether a particular configuration is actually the one which best fits the data, i.e. that a 'global minimum' has been reached. But there are, at least, ways of guarding against entrapment in a local

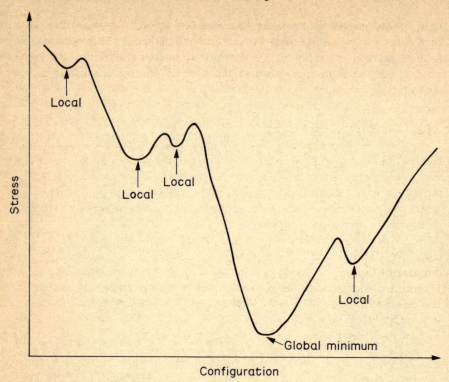

Figure 3.10   *Local and global minima*

minimum, and fairly reliable ways of detecting a local minimum:

(a)   Reliable modern computer programs avoid starting from an *arbitrary* initial configuration, and produce an initial configuration which is likely to be fairly close to the global minimum (see Appendix A3.2). This is like using what information we have about the terrain to position our climber close to where we think the lowest valley is.

(b)   Programs use more than one minimising procedure to capitalise on the strengths of each approach. (Thus MINISSA uses the somewhat erratic but rapid Guttman technique at the start and then switches to the smoother Kruskal technique when the earlier phase shows no signs of systematic improvement.)

(c)   It is always sensible to obtain MDS solutions in a number of different dimensions (e.g. in five through one dimensions, for reasons discussed in section 3.7.1 below). The stress value should *decrease* as the number of dimensions increases. If it actually increases then that solution is bound to be a local minimum.

(d)   The best safeguard against local minima, and one which is strongly recommended, is to use *several* different starting configurations (implemented in MINISSA by using the READ CONFIG command) and see whether they produce markedly different results. (Do not rely upon just looking at the resulting configurations: remember that a reflected and/or rotated configuration can *look* very different to the original one, though be identical to it in the sense of being a legitimate similarity transformation. If in doubt it is safest to use PINDIS to compare configurations; see 7.3.2 below).

(viii) **Move the old configuration to a new positioning of points.** When all the correction factors and the step size have been calculated, the general correction formula is used, and every point is moved into greater conformity with the data. Thus a new configuration has then been obtained and the iterative cycle is complete.

3.5.5 The final configuration

In the basic model, and indeed in most MDS models, the final configuration is rotated to principal components before being output. This rotation does not in any way change the pattern of points or the relative distances between them, but principal components (or 'principal axes') provide a framework of reference axes which possesses some useful statistical properties. It may be helpful to recapitulate the method of principal components.

Given a set of points located in a multidimensional space (in this case the final configuration) the method of principal components first finds the line (axis dimension) through the configuration which has maximum variation, i.e. along which the co-ordinates of the points are maximally spread or differentiated. This line is termed the first 'principal component' or 'principal axis' of the space. Following this a second axis is found which is orthogonal to the first axis, i.e. which is statistically independent of the first component, in the sense that the correlation of the co-ordinates of the points on the two components is zero, and also explains the maximum amount of the remaining variation. This is the second principal component. The process continues in this manner—identifying axes which are orthogonal to those already found and which explain maximum amounts of remaining variation—until the final components are normally explaining trivially small proportions of the total variation. In this sense, principal components is often viewed as a way of orienting the configuration so that variation is concentrated into as few dimensions as possible.

The actual amount of variation represented by a particular component is given by the size of its latent root (also called the eigenvalue) which can be thought of as a standard deviation measuring the dispersion of the objects along that dimension (and indicated by the sigma value at the foot of each column of the final configuration printout in some MDS(X) programs). Comparison of the values of sigma is often instructive: the more equal they are, the more circular (or spherical in three-dimensions) the pattern of points in the final configuration is. Conversely, the more unequal they are, the more ellipse-like the pattern is. If any sigma values are close to zero, this signals the fact that there is virtually no variation on the dimension concerned—i.e. that the dimensionality chosen for analysis is unnecessarily high.

3.6 Assessing the Solution

We now want to illustrate the process of finding a 'best solution' to a set of data using non-metric MDS. Information from the process is used to help decide how good a solution has been obtained and diagnose inadequacies in it. This will be done by using a genuine set of data, as opposed to one chosen for purely illustrative purposes.

Table 3.4 Co-occurrence frequencies of sortings of occupational titles

(BC)

	CA	SST	GM	EM	ST	SW	C	AD	CPR	MOR	PL	MPN	BCK	PST	UMO	PM	CE	PHT	BSL	RCK	AP	A	RED	PO	GEO	SMG	TDH	TDR	ESG	JN	LT
SST	25																														
GM	3	2																													
EM	3	2	9																												
ST	3	29	6	2																											
SW	43	40	61	2	17																										
C	13	1	21	11	3	5																									
AD	2	4	8	21	2	12	19																								
CPR	2	17	5	5	32	18	6	4																							
MOR	27	43	3	2	20	39	7	17	7																						
PL	26	2	58	10	3	6	12	7	12	4																					
MPN	4	22	13	5	9	39	20	22	7	25	13																				
BCK	10	16	10	8	23	26	19	25	33	11	10	21																			
PST	22	54	3	2	20	47	11	8	21	36	3	33	27																		
UMO	15	1	17	39	1	3	33	22	3	3	15	8	7	7																	
PM	1	16	10	12	34	29	2	25	16	3	9	12	29	9	5																
CE	6	27	4	1	8	7	3	1	17	17	8	28	16	14	5	14															
PHT	45	4	11	11	2	8	17	8	6	9	6	3	12	5	1	3	7														
BSL	5	1	15	42	2	17	1	20	18	19	18	11	5	1	63	6	20	35													
RCK	1	36	42	28	2	20	20	35	4	6	9	8	16	5	18	18	7	3	22												
AP	4	14	28	3	9	37	37	15	17	35	7	8	9	8	4	4	31	9	4	11											
A	26	8	10	6	15	11	8	2	14	7	3	9	7	9	11	20	10	35	3	28	7										
RED	10	8	3	12	12	12	9	4	1	3	33	5	13	5	9	5	3	18	13	22	13	18									
PO	2	13	5	40	7	1	7	3	19	4	4	1	11	1	34	14	10	8	35	6	20	8	6								
GEO	2	18	9	38	10	6	3	5	8	26	19	8	26	7	2	2	2	9	26	3	11	10	6	14							
SMG	37	1	7	35	3	13	2	2	19	15	9	12	9	1	8	8	8	8	57	5	25	39	17	34	3						
TDH	25	1	16	0	26	2	18	4	5	1	9	11	7	56	2	1	2	9	2	20	11	3	32	36	3	6					
TDR	2	28	18	3	3	18	19	3	1	19	7	8	5	8	1	7	3	14	1	23	8	8	17	2	4	2	1				
ESG	2	17	3	0	5	3	8	5	14	9	14	3	9	3	25	3	34	37	3	10	13	10	12	2	10	6	4	3			
JN	33	8	9	3	21	19	4	2	13	12	25	1	5	15	22	15	19	14	1	30	3	24	18	18	34	24	4	3	1		
LT	11	2	23	39	19	8	17	4	35	27	19	16	7	20	11	20	19	6	38	19	10	8	18	4	18	11	6	8	13	12	
BC	12	14	3	3	12	4	2	12	10	9	4	3	9	37	15	1	9	4	18	2	23	51	14	5	2	38	44	1	10	3	5

3.6.1 An illustration of the iterative procedure in non-metric MDS

In a study (Coxon and Jones 1978a, p. 42 et seq.) of the natural groupings which people use to classify occupations, a group of 71 individuals were asked to sort a set of 32 occupational titles into as many or as few groups as they wished. A measure of pairwise similarity between the occupations was defined as the frequency with which two occupations were sorted into the same group: the greater the number of subjects who put a pair of occupations together, the more similar they are defined to be. (This co-occurrence measure is *M1*, discussed earlier in 2.2.3.3). The frequency matrix is presented in Table 3.4. The data are analysed by the basic non-metric distance model in two dimensions, using stress$_2$ and Kruskal's weak monotonicity.

In order to dramatise the process of improvement, a deliberately poor starting configuration was chosen (stress$_2$ = 0.986). The iterative process is now examined at the 2nd, 5th, 10th and 23rd iteration to see what progress is made.

Iterations 0–2 (Fig. 3.11)
Figure 3.11a shows the moves made in the positioning of the points from the initial configuration (iteration 0) to the second iteration. Note that there is very little change in positioning, but that the step size increases. By iteration 2, the monotone fitting function (Figure 3.11b) is beginning to descend from the top left to the bottom right of the Shepard diagram (since similarities are inversely related to distances). The transformation function is little better than a straight vertical line, indicating that a very large number of data values are being fitted by the same disparity value.

Iterations 2–5 (Fig. 3.12)
Clearly, the positioning of the points is changing fast (Figure 3.12a) and improving rapidly, with a reduction in the size of stress$_2$ by a third. Note that larger step sizes are occurring. By iteration 5, the monotone fitting function (Figure 3.12b) is taking on its characteristic 'steppy' form and the larger number of steps at the bottom right hand of the function shows that the improvement is concentrated principally in the positioning of the smaller similarity values.

Iterations 5–10 (Fig. 3.13)
There is a dramatic improvement from iterations 5 to 6 (Figure 3.13a) but beyond here it tails off somewhat, and the step size decreases fairly drastically. In terms of improvement, the point of diminishing returns is setting in. The monotone function is now fitting the highest and lowest similarities well, though the middling data values are still not well fit. Note that monotone function is taking on a recognisably convex form (cf. Shepard 1974, p. 400), suggesting that the co-occurrence frequency data might well be related to distance in a more regular (polynomial) manner.

Iterations 10–23 (Fig. 3.14)
The improvements are now very slight indeed; stress$_2$ decreases by the same amount between iterations 9 and 11 as between 11 and 23. Also, the step size becomes smaller and smaller, till the process finishes at iteration 23. The Shepard diagram (Figure 3.14b) for iteration 23 is virtually identical to that for iteration 10, indicating some slight improvement.

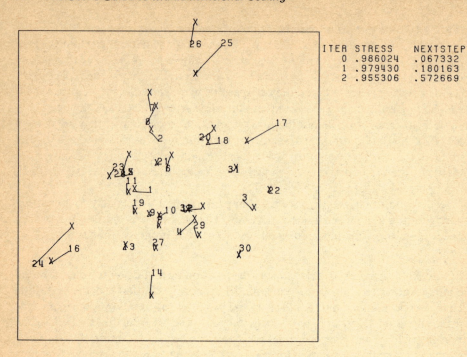

```
ITER STRESS    NEXTSTEP
  0  .986024    .067332
  1  .979430    .180163
  2  .955306    .572669
```

Figure 3.11a   *Iterations 0–2: moves in positioning of points*

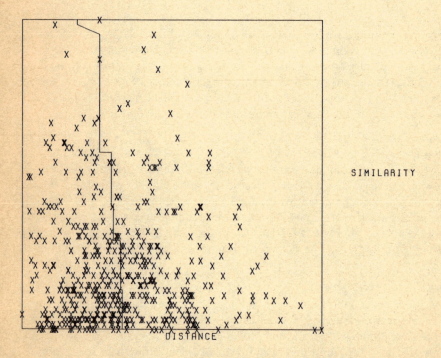

Figure 3.11b   *Shepard diagram at iteration 2*

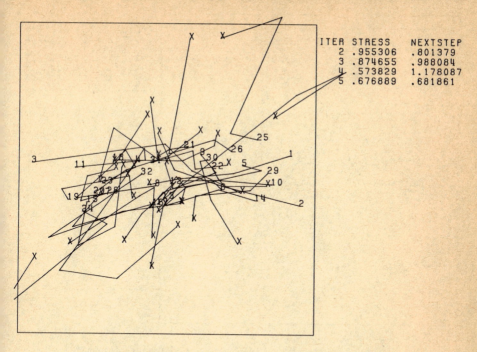

Figure 3.12a *Iterations 2-5: moves in positioning of points*

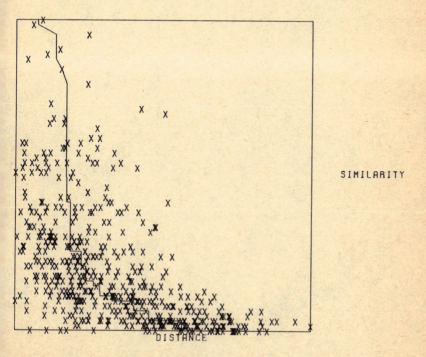

Figure 3.12b *Shepard diagram at iteration 5*

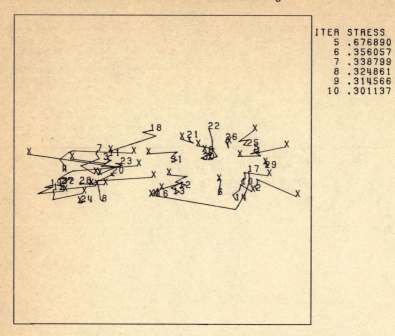

ITER	STRESS	NEXTSTEP
5	.676890	.516218
6	.356057	.414221
7	.338799	.102525
8	.324861	.090250
9	.314566	.066588
10	.301137	.064490

Figure 3.13a   *Iterations 5–10: moves in positioning of points*

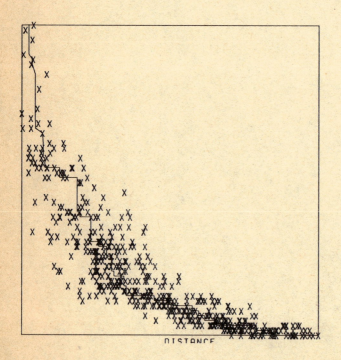

SIMILARITY

Figure 3.13b   *Shepard diagram at iteration 10*

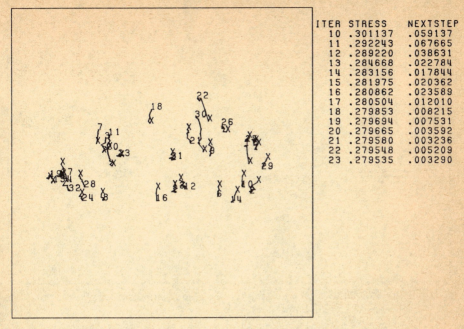

ITER	STRESS	NEXTSTEP
10	.301137	.059137
11	.292243	.067665
12	.289220	.038631
13	.284668	.022784
14	.283156	.017844
15	.281975	.020362
16	.280862	.023589
17	.280504	.012010
18	.279853	.008215
19	.279694	.007531
20	.279665	.003592
21	.279580	.003236
22	.279548	.005209
23	.279535	.003290

MINIMUM ACHIEVED

Figure 3.14a   *Iterations 10–23: moves in positioning of points*

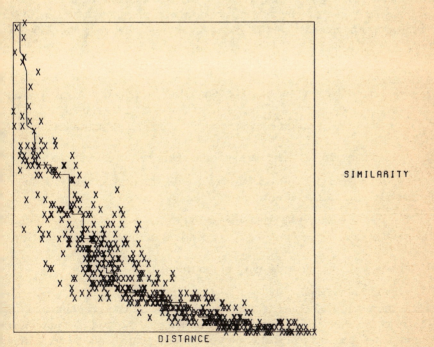

SIMILARITY

DISTANCE

Figure 3.14b   *Shepard diagram at iteration 23*

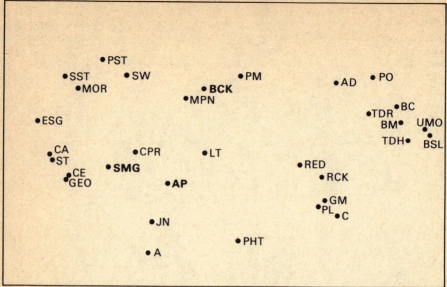

Stress$_2$ ($\hat{d}$) = 0.281

Figure 3.15   *2-D MDS final configuration (rotated to principal axes): co-occurrence of 32 occupational titles*

After the final, 23rd iteration, the configuration is rotated to principal axes and output, together with information on fit (stress values) and, if desired, also on the distances, the disparities and the residuals ($d_{jk} - d_{jk}^0$). The final configuration (centred and normalised) is given in Table 3.5, and is plotted in Figure 3.15. (Since the axes may be rotated at will, they are not drawn in; this allows attention to be concentrated upon characteristics of the configuration other than the arbitrary positioning of the axes.)

3.6.2 Diagnostics

Before beginning to interpret the final configuration, it is a good practice to assess first the adequacy and stability of the solution. The main questions are:

   (a)   Is a configuration with this stress value an acceptable approximation to the data?

   (b)   Is the dimensionality we have chosen likely to be the correct one?

(The answers to these questions are deferred until the next section.) Of more immediate relevance is the following set of questions:

   (c)   How well is each datum fit by the solution?
Are the residual values evenly distributed, or concentrated in a few rather badly fit instances?
Is one particular stimulus causing most of the trouble?

3.6.2.1 Analysis of residuals

The basic information necessary to answer all these questions is contained in the final Shepard diagram (see Figure 3.14b), which should be read like any other regression plot. It consists of a scatter plot of the data points with the best-fitting transformation function (in this case, the weak monotonic or ordinal function) drawn through it. The horizontal spread from this function represents the overall

Occupational Titles			Final Configuration Dimensions	
Abbrev.	No.	Title	I	II
CA	1	Chartered accountant	−1.246	−0.082
SST	2	Secondary school teacher	−1.152	0.472
GM	3	Garage mechanic	0.733	−0.392
BM	4	Barman	1.262	0.156
ST	5	Statistician	−1.241	−0.117
SW	6	Social worker	−0.703	0.488
C	7	Carpenter	0.814	−0.496
AD	8	Ambulance driver	0.804	0.416
CPR	9	Computer programmer	−0.638	−0.061
MOR	10	Minister of religion	−1.045	0.391
PL	11	Plumber	0.678	−0.431
MPN	12	Male psychiatric nurse	−0.267	0.321
BCK	13	Bank clerk	−0.144	0.391
PST	14	Primary school teacher	−0.886	0.574
UMO	15	Unskilled machine operator on a factory assembly line	1.445	0.110
PM	16	Policeman	0.108	0.473
CE	17	Civil engineer	−1.126	−0.229
PHT	18	Photographer	0.094	−0.678
BSL	19	Building-site labourer	1.485	0.049
RCK	20	Restaurant cook	0.704	−0.227
AP	21	Airline pilot	−0.399	−0.285
A	22	Actor	−0.556	−0.785
RED	23	Railway engine driver	0.555	−0.149
PO	24	Postman	1.078	0.417
GEO	25	Geologist	−1.143	−0.254
SMG	26	Sales manager	−0.827	−0.154
TDH	27	Trawler deckhand	1.335	0.033
TDR	28	Taxi driver	1.049	0.221
ESG	29	Eye surgeon	−1.353	0.156
JN	30	Journalist	−0.528	−0.543
LT	31	Laboratory technician	−0.138	−0.059
BCR	32	Bus conductor	1.249	0.275
		Mean	0.000	0.000
		S.D.	27.762	4.238

$Stress_2 (\hat{d}) = 0.2808$ (primary approaches to ties)

Table 3.5   *Final 2-dimensional configuration of the data from Table 3.4*

degree of fit (recall that stress is a normalised measure of the dispersion of the distances from this monotonic regression function). The further a data point is from its corresponding disparity value (measured on the distance axis) the worse fit it is, and the larger the associated residual value ($d_{jk} - \hat{d}_{jk}$) will be. The matrix of residual values, rounded and multiplied by 10 (so that 0 represents a residual between 0 and 0.499, 1 a residual between 0.5 and 1.499 etc.) is presented in Table 3.6, accompanied by a frequency diagram. As is often the case, the distribution of residuals is strongly skewed toward the small values, indicating overall goodness of fit. But it is worthwhile giving special attention to the high residual values, and especially to the eighteen with values over 0.45 (i.e. values of 5 and 6 in Table 3.6). A

Table 3.6 — Absolute values of residuals $(d_{jk} - \hat{d}_{jk})$, ×10, rounded

		(1)	(2)	(3)	(4)	(5)	(6)	(7)	(8)	(9)	(10)	(11)	(12)	(13)	(14)	(15)	(16)	(17)	(18)	(19)	(20)	(21)	(22)	(23)	(24)	(25)	(26)	(27)	(28)	(29)	(30)	(31)	(32)
CA	1	—																															
SST	2	0	—																														
GM	3	0	2	—																													
BM	4	3	0	3	—																												
ST	5	0	2	1	1	—																											
SW	6	2	0	0	1	1	—																										
C	7	2	0	2	0	0	0	—																									
AD	8	1	1	2	0	2	1	5	—																								
CPR	9	0	0	1	0	1	2	2	1	—																							
MOR	10	0	0	0	0	0	0	0	1	0	—																						
PL	11	2	2	0	0	2	2	0	5	0	3	—																					
MPN	12	0	2	2	1	2	1	0	2	0	3	2	—																				
BCK	13	5	1	1	0	2	0	4	0	0	2	1	6	—																			
PST	14	0	0	0	1	1	0	0	0	1	3	0	0	2	—																		
UMO	15	0	0	0	0	0	5	2	2	3	0	1	1	0	3	—																	
PM	16	4	2	0	1	3	0	0	0	0	0	0	2	2	0	3	—																
CE	17	1	0	1	0	2	0	0	3	2	3	0	5	4	0	0	4	—															
PHT	18	2	2	2	4	0	0	1	0	0	2	2	2	3	0	4	0	4	—														
BSL	19	0	0	0	1	0	3	2	0	3	0	3	2	0	0	2	0	2	5	—													
RCK	20	3	1	4	2	3	2	0	0	0	0	1	0	0	0	4	0	2	2	2	—												
AP	21	3	3	0	1	2	0	1	1	1	2	3	1	1	0	0	1	0	0	4	1	—											
A	22	0	1	2	1	0	3	2	0	3	0	2	3	2	0	2	3	0	2	0	4	5	—										
RED	23	1	0	4	0	2	1	1	3	1	0	0	0	0	2	0	0	0	0	0	0	2	2	—									
PO	24	5	0	3	0	0	2	0	3	0	1	3	1	0	0	1	0	0	0	4	1	2	0	0	—								
GEO	25	2	5	1	1	1	3	4	0	0	2	2	4	2	1	2	1	0	1	0	1	0	2	1	3	—							
SMG	26	1	1	3	0	2	1	4	0	0	3	3	0	1	2	4	3	2	3	0	0	5	2	2	0	5	—						
TDH	27	2	3	1	0	0	2	3	1	2	2	1	3	3	0	0	0	0	0	0	3	3	0	0	4	0	3	—					
TDR	28	2	0	5	1	0	2	0	0	3	0	0	1	2	2	0	2	5	2	0	3	0	0	0	1	2	1	0	—				
ESG	29	2	2	3	0	0	0	2	3	2	4	2	2	2	3	0	4	2	5	2	4	6	0	2	2	4	2	2	2	—			
JN	30	3	2	0	1	2	0	2	2	0	0	0	0	1	2	1	0	0	0	0	0	1	4	0	3	1	1	1	4	3	—		
LT	31	1	3	2	1	3	1	1	1	3	1	3	2	0	1	2	1	2	0	0	2	4	0	1	1	2	4	1	0	0	0	—	
BCR	32	0	1	0	0	1	0	3	0	0	3	0	3	1	3	0	3	2	1	5	2	1	5	3	3	4	0	4	3	3	3	3	—

Table 3.6 *Absolute values of residuals* $(d_{jk} - \hat{d}_{jk})$, ×10, rounded

look along row, and down column, 21 of Table 3.6 (corresponding to the Airline Pilot) shows that seven of the worst fit values occur in association with this one occupational title. A similar inspection will show that 26 (Sales Manager) and 13 (Bank Clerk) are also over-represented among the highest residuals.

A more systematic way of examining where badness of fit is concentrated is to look at the contribution which each point makes to the overall stress value, and this is done in Figure 3.16.† Note that we are taking into account all the residuals in which each point is involved, and not just the extreme ones. Nonetheless, the same conclusion is evident: points 21 and 26 contribute significantly more than the others to the badness of fit. We do not know why these particular points should be so troublesome, but analysis presented elsewhere (Coxon and Jones 1979, p. 42 et seq.) suggests that the worst fit occupations have characteristics which are not common to the remaining ones. In any event, it is worth considering simplifying the analysis either by removing the worst fit point(s) by deleting the relevant rows and columns from the data matrix and re-running the program, or by re-running the program in a higher dimensionality to see whether the additional dimensions allow the fit to be significantly improved.

It could also be that we have encountered a local minimum, in the sense that some other configuration may exist with lower stress, which would locate these worst-offending points in another position but would be substantially similar in other respects. The only way to check this,‡ is to re-run the program with a different initial configuration—say, the current final configuration with the points relocated to where the user thinks they ought to be. In the present example, different starting configurations and methods of minimisation produce virtually identical configurations (compare Figure 3.15 with Figure 2.8 in Coxon and Jones, 1978a, p. 43) so we can conclude with a fair degree of certainty that the point locations are as accurate as they can be. Nonetheless, the actual location of the worst-fit points must be treated with considerable caution when interpreting the configuration, and to emphasise this, these points are in bold print in Figure 3.15. (We shall return to the interpretation of Figure 3.15 in the next chapter.)

3.6.3 Degenerate and trivial solutions

Occasionally, final configurations can be produced which have very low stress values but are substantively meaningless. This arises when the low stress value has been obtained by the program capitalising on some technical feature of the minimisation process, such as weak monotonicity, or upon some unanticipated features of the data. Such configurations are often termed 'degenerate' or 'trivial' solutions (see Shepard 1974, pp. 391–9).

A good example of this occurs in very highly clustered data. If the data are such that the stimuli fall into a small number of clusters where the dissimilarities within each cluster are uniformly smaller than those between the clusters, the effect will be to produce a very low stress solution, where points within a cluster will condense or

---

†The individual point contributions do not add up to the stress value, because we are actually examining all the $(p - 1)$ *pairwise* contributions to stress involving a particular point. The point contributions do not therefore contribute additively to stress.

‡Interactive MDS programs such as SPACES (Schneider and Weisberg, 1974) allow the user to re-position or delete points to see how stress values change.

Stress point	(No)
0.1148 AP	21
0.1035 SMG	26
0.0904 BCK	13
0.0856 MPN	12
0.0822 CE	17
0.0816 A	22
0.0785 LT	31
0.0778 JN	30
0.0776 ESG	29
0.0727 ST	5
0.0716 PM	16
0.0694 CPR	9
0.0686 GEO	25
0.0670 BM	4
0.0669 RCK	20
0.0666 PHT	18
0.0653 TDR	28
0.0644 AD	8
0.0636 SST	2
0.0633 PL	11
0.0619 PO	24
0.0613 BC	32
0.0613 MOR	10
0.0604 SW	6
0.0560 C	7
0.0553 CA	1
0.0545 TDH	27
0.0528 RED	23
0.0526 GM	3
0.0525 PST	14
0.0497 BSL	19
0.0434 UMO	15

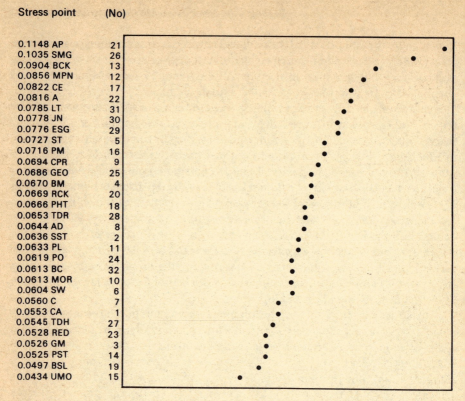

Figure 3.16   *Individual point contributions to total STRESS, of 0.2808 in two dimensions*

even collapse upon a single position and the program will then be able to minimise stress simply by maximising the distance between the positions of the clusters. This is not to say that if there are recognisable clusters of points in a final configuration the solution is therefore degenerate. For instance, it is not difficult in Figure 3.15 to spot at least two fairly coherent, genuinely distinct, clusters at the right hand side of the space, the groupings (TDR, BC, BM, UMO, TDH, BSL) and (RED, RCK, GM, PL, C), which comprise the 'unskilled' and 'trades' categories used by the subjects who made the judgments on which the data are based (Coxon and Jones 1979, pp. 39–41). However, if a degenerate clustering does occur, it is worthwhile making a separate scaling analysis of the stimuli involved in each cluster.

   Two other examples of possible degeneracy have already been mentioned earlier when discussing types of monotonicity and different approaches to ties in the data. First, *weak* monotonicity allows distinct data values to be fit by the same disparity value; indeed, the block-averaging procedure used in monotone regression does just this. Usually there should not be a markedly smaller number of blocks (disparity values) than there are distinct data (dissimilarity values). However, if there are very few blocks and they contain a large number of entries, then the solution may well be degenerate. This situation shows up on the Shepard diagram in the form of long vertical segments on the monotone function, each with a large number of associated data points, because a large number of data are being fitted by a single disparity value.

Secondly, if the primary approach to ties is chosen, the program is given the freedom to fit different disparity values to data which have the same value, without this counting towards badness of fit. Sometimes this freedom is grossly exploited by the program, especially when the data contain only a few distinct values (for example, when a 5-point rating scale is used on pairwise judgments of a large number of stimuli). Where this occurs it is indicated on the Shepard diagram by the appearance of long horizontal segments on the monotone function, highly populated by data points.

It is very difficult to determine decisively whether a final configuration is 'really' degenerate or trivial, but this is not the point. Rather, the user should be alerted to the danger signs of artificially low values of stress which can often indicate serious loss of information, and to the reasons for their occurrence. At the very least, it is good practice always to inspect the Shepard diagram and the set of disparity values relating to the final configuration. If any of the tell-tale signs of a trivial or degenerate solution appear in a given solution, then a re-analysis should be made using options which counteract the weakness concerned.

1   If the researcher has objective information, or even a strong hunch, about what the configuration *should* look like, then Confirmatory MDS should be used, or alternatively the points whose positioning is known should be fixed (by the FIX POINTS command in SSAM or by using PREFMAP), and the analysis run to determine the position of the other points.

2   If the monotone function is very 'steppy', containing many long vertical and horizontal segments, then a more restrictive function which excludes such steps may be used in preference to the monotone function. A fairly common option is to choose to fit a *linear* (or even a power) function between the data and the distances, and this can be done by using the MRSCAL (*metric scal*ing) program (see 6.1.4).

3   If the difficulties arise because the primary approach to ties has been used, the data can be re-analysed with MINISSA, using the *secondary* ties option. If they arise due to weak monotonicity, the data can be re-analysed using SSAM, applying the strong monotonicity option.

Having discussed the relatively rare and unusual problems of degeneracy, let us turn to the more important general issue of assessing the stress value of a configuration.

3.7   Stress, Dimensionality and Recovery of Metric Information
Three important and related issues arise in using MDS programs. These are:

(i)   How is the stress value of a configuration to be interpreted?   .

(ii)   What is the 'real' or 'best' dimensionality for a solution?

(iii)   How well can non-metric MDS recover information if the data are 'noisy' or error-prone?

3.7.1   Evaluation of stress
A number of factors affect the value of stress. The most important are:

(i)   *The number of points.*   In general, the larger the number of points the more the information to be fitted and the higher the stress.

(ii)   *The dimensionality of the solution.*   The higher the dimensionality of the solution, the easier it is to fit the information, and therefore the lower the stress value. In general, it is possible to fit *any* data relating to p points perfectly in $p - 2$ dimensions. Such a solution is termed a 'trivial solution'.

As we have noted before, a further set of factors holds as a necessary result of options (or default values) chosen by the user in obtaining a solution, and discussed above. These are:

(iii)   *The type of stress.*   Raw stress/phi is necessarily larger than normalised stress and, because of the different normalising factors, $stress_2$ is normally about twice as large as $stress_1$.

(iv)   *The type of monotonicity criterion.*   Guttman's strong monotonicity criterion is more stringent than Kruskal's weak monotonicity criterion and therefore stress based on strong monotonicity will necessarily be larger than stress based on weak monotonicity.

(v)   *Tying approach.*   The secondary approach to ties, treating tied values as equivalences, produces higher stress values than the primary approach.

Because stress is affected by all these factors, it is meaningless to talk about what an 'acceptable value of stress is' without further specification. To simplify matters, we shall therefore restrict attention to the paradigm case of stress:

stressform 1;
based upon Kruskal's $\hat{d}$ (weak monotone) fitting quantities;
using the primary approach towards ties.

In asking what is an acceptable level of stress, we are asking a variant of the common statistical question, 'Does the non-metric MDS model fit the data well enough that the stress value could not have arisen by chance?' (Cf. Kruskal 1972b.) Put slightly differently, we advance the null hypothesis that some chance mechanism could have generated the data, and we use this as a bench-mark to assess how far the actual configuration departs from a random distribution. Very little progress has been made in analytically deriving a statistical distribution of stress, and recourse has been had instead to so-called Monte Carlo simulation methods (Young 1970; Wagenaar and Padmos 1971; Isaac and Poor 1974; Spence 1972; Spence and Graef 1974).

3.7.2   Simulation of stress values

The basic process in simulating the distribution of stress consists of producing a configuration of p points in t dimensions and calculating the distances between the points in the configuration. This is the 'true' or generating configuration. The distances (or in some cases the coordinates) are then distorted by adding to them 'noise', i.e. error in differing but specified amounts. Sometimes in addition the distances are transformed monotonically. The resulting set of error-perturbed distances are now treated as if they were dissimilarities data, and are scaled by a non-metric MDS distance model program in a number of different dimensions, including the 'true' dimensionality t (for we wish to assess the effect of scaling data in the wrong dimensionality). A large number of such dissimilarity matrices are generated which vary in the number of points in the configuration, the 'true' dimensionality and the amount of error added. All of these matrices are scaled and the resulting stress$_1$ values are noted.

Spence and Graef's M-SPACE *procedure*

Although there are several variants of the procedure, the most well known is the Spence-Graef M-SPACE procedure (Spence and Graef 1974), a variant of which is incorporated in the MINISSA program in the MDS(X) series as a way of helping users decide upon the 'true' dimensionality and likely error present in their data.

Spence and Graef constructed random configurations containing a given number of points ($p = 12, 18, 26, 36$) in a given number of dimensions ($t = 1, 2, 3, 4, 5$), and then added error at five levels (from a unit normal distribution with standard deviations σ, of $0, 0.0625, 0.1225, 0.2550$ and infinity) to each co-ordinate in this case. For each combination of points, true dimensionality and error level, a number of dissimilarity matrices was produced and these were scaled in five through one dimensions. The resulting stress$_1$ values were averaged and put together to produce a set of diagrams (nomographs) such as those reproduced in Figure 3.17.

This set of four diagrams refers to configurations of 36 points with *true* (generating) dimensionality of 1 (top left hand diagram), 2 (top right), 3 (bottom left) and 4 (bottom right). (Similar diagrams exist for from 12 to 36 points). Within each diagram there are five 'curves'—one for each dimensionality m in which the configurations were actually scaled. Each curve joins the average stress$_1$ values of the configurations as the level of added error is increased. For example, (see the top right hand diagram) in a true dimensionality of 2 and with no added error, the configurations of 36 points scaled in two, three, four and five dimensions yield a

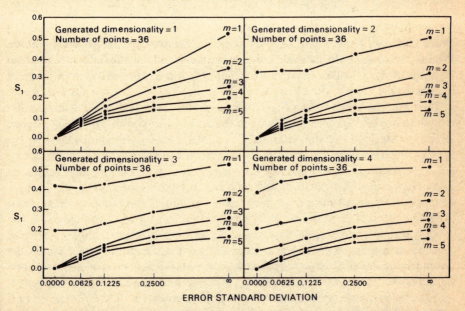

Figure 3.17 *Stress of a recovered configuration in spaces of varying dimensionality as a function of the error level in a known undelying configuration*

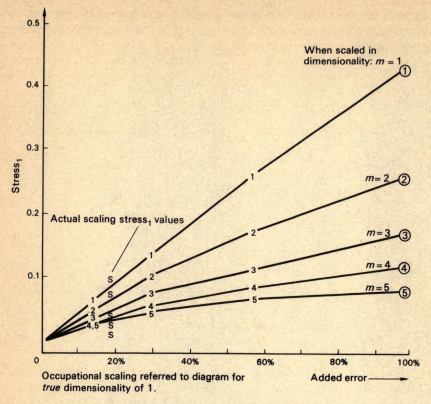

Figure 3.18   *Spence-Graef M-SPACE analysis*

zero-stress$_1$ solution, whilst those scaled in *one* dimension have a stress$_1$ value of about 0.32 on average. At each level of error, the stress$_1$ of the solutions in two through five-dimensions become more separated—and it is the difference between these values which will be important when the diagrams are used to interpret actual scaling results.

The idea of M-SPACE is simply to compare the actual set of stress$_1$ values which the user has obtained with the results of the Spence-Graef simulations, in order to assess the likely 'true' dimensionality, and the probable amount of error. An example will help illustrate the process. Coxon and Jones (1979, pp. 73–4) scaled a set of averaged pairwise similarity ratings referring to 16 occupations, obtaining stress$_1$ values of: 0.024, 0.030, 0.043, 0.060 and 0.100 for solutions in 5, 4, 3, 2 and 1 dimensions. This 'actual set' of stress values was then compared to Spence and Graef's results obtained from scaling random configurations of 16 points. The M-SPACE procedure compares the actual set of stress$_1$ values with each 'true dimensionality' diagram in turn, finding the point at which the actual set fits the simulated results most closely by locating the point along the error axis at which the actual stress$_1$ values conform most closely to the simulated ones. As a result, the user is given the error level to which the actual set of stress$_1$ values best corresponds by reference to each 'true' dimensionality in turn. Usually the fit will be best in one particular 'true' dimensionality, and this information is taken to mean that this is the most likely underlying dimensionality of the user's data. To our surprise, the use of M-SPACE on these data strongly suggested a 'true' one-dimensional (or possibly a two-dimensional) solution and the relevant M-SPACE nomograph for true dimensionality of 1 is presented in Figure 3.18: the 'actual set' fits best the true dimensionality of 1 for an error level of 18 per cent, (on any account a low level of error, and by far the best fitting in any of the true dimensionalities). M-SPACE should not be used uncritically as an automatic detector of 'real' dimensionality: if anything, it tends to underestimate dimensionality, and the authors mention a number of other cautions (Spence and Graef 1974, p. 3).

The results of other simulation studies point in the same direction and confirm Shepard's (1966) intuitions about the number of points and dimensions needed to produce a stable solution. Klahr (1969) shows, for example, that stress values of sets of randomly generated dissimilarities for between 6 and 8 points in three-dimensions yield 'good' solutions according to Kruskal's original rules! Fortunately, a small increase in points rapidly diminishes the likelihood of such a mistaken inference. Stenson and Knoll's (1969) study is interesting mostly for the evidence it provides of the effect on $stress_1$ of the choice of primary or secondary approach to ties. To estimate this, he ties dissimilarity values for 30 points 'coarsely' (into 10 approximately equal sets of tied values) and 'finely' (into 50 such sets). Surprisingly the fineness of grouping has little apparent effect on $stress_1$ values. Wagenaar and Padmos (1971) were the first to provide a realistic and systematic investigation of the effect of adding error into the process of generating the dissimilarity matrices, and their results here have been generalised by Spence and Graef.

3.7.3 Other approaches to assessing dimensionality, stability and metric recovery

From the practice and lore of factor analysis come other approaches to determine the likely dimensionality of the data. The most famous is the 'scree test', or 'elbow test'. The recommended strategy here is to perform the scalings in a high number of dimensionalities, stepping down to a uni-dimensional scaling. The stress values are then charted against the dimensionality, and joined together to form a polygon. If a noticeable bend or 'elbow' occurs, indicating that the improvement in fit is not significantly altered by the addition of a further dimension, then the lower-dimensional solution is to be preferred. A common variant of this rule is ascribed (probably unfairly) to Shepard and often referred to as 'Shepard's Law': if a solution exists, it probably exists in two dimensions. Extended a little, such a rule contains a good deal of sense. Uni-dimensional solutions are quite often degenerate and unable to portray a situation that sometimes occurs—that the points in fact form a non-linear, but uni-dimensional sequence such as a 'horseshoe' (see Kendall 1971a, pp. 224–7 and 4.6 below). So even if a uni-dimensional solution is likely, it is prudent to scale in two dimensions. A two-dimensional solution is the easiest to comprehend because it can be readily assimilated, and whereas a third dimension can be relatively easily visualised, higher dimensionalities, can not. 'If a solution exists, it probably exists in two dimensions; if it doesn't then it certainly exists in three', might be a reasonable extension to Shepard's law.

As we have seen, Shepard (1966, p. 288) shows that non-metric constraints, if imposed in sufficient number, begin to act like metric constraints. That is, information on the pairwise ordering of stimuli is normally sufficient to constrain the location of points in a configuration to such an extent that the distances between them are virtually fixed. The key word is 'normally'. In fact, the number of data need to exceed the number of points times the number of dimensions to a considerable degree before the configuration is really stable. Again, Monte Carlo studies have been used extensively to see how well non-metric scaling can recover *known* configurations (see especially Young 1970 and Shepard 1966). The results are conclusive: so long as the number of order relations in the data exceed the number of co-ordinates of the solution by at least a factor of 2 (and so long, obviously, as there is not an overwhelming amount of error), non-metric analysis can recover the correct underlying configuration (or, rather, the distances) extremely accurately. For 'true' dimensionality of two, the extent of metric recovery has been studied extensively, and the results of Shepard (1966, p. 299) still hold:

> While the reconstruction of the configuration can occasionally be quite good for a small number of points, it is apt to be rather poor (for p less than eight, say). As p increases, however, the accuracy of the reconstruction systematically improves until even the worst of ten solutions becomes quite satisfactory with 10 points, and, to all practical purposes, essentially perfect with 15 or more points.

But, as he later points out (Shepard 1974, p. 395), the lesson has not always been well heeded by users:

> A distressing number of two- and even three-dimensional solutions have been published in which, despite the inclusion of only six to eight objects, no evidence is provided that the configuration has a reasonable degree of metric determinacy and is not a prematurely arrested case of convergence toward a degeneracy.

A footnote should be added to these studies of recoverability and stability of MDS solutions which is of particular importance to users who wish to have subjects judge a large number of stimuli, and is also of general importance in interpreting a configuration:

> Information about *larger* distances/dissimilarities is far more crucial in ensuring satisfactory recovery of metric information than is information about medium or smaller dissimilarities; and
> small distances in the solution are far less stable than large ones, implying that 'global' information (between distant points or clusters of points) is a more reliable basis for interpreting a solution than 'local' information between highly proximate points.

These issues are given extended treatment in an important study by Graef and Spence (1979).

APPENDIX A3.1 COMPARISON OF MEASURES OF FIT BETWEEN DATA AND SOLUTION USED IN NON-METRIC MDS

1   All measures of fit used in the basic model compare a set of disparities (ratio-level quantities which are a function of the data, $d_{jk}^0 = f(\delta_{jk})$, which are *monotonic* for ordinal data and *linear* for metric or interval data with the ratio-level distances, d_{jk}. Two forms of comparison are used:

(i)   *the difference* $(d_{jk} - d_{jk}^0)$, which forms the basis of badness-of-fit measures, since the greater the discrepancy between a solution and the data, the greater will be the differences; and

(ii)   *the scalar product* $(d_{jk}d_{jk}^0)$, which forms the basis of goodness-of-fit measures, since the greater the covariance (or the less the angular separation) between data and the solution, the greater will be the scalar products.

2   The basic measure of goodness-of-fit used in non-metric programs emanating from the Bell Laboratories is *stress*, and in particular (normalised) stress$_1$, based upon Kruskal's BFMF disparities, $\hat{d}_{jk}$. These and alternative measures are dealt with in sections 3.3 and 3.5.2 and are extensively discussed in Kruskal and Carroll, 1969.

3   The measures used by Lingoes (Michigan), Guttman (Israel) and Roskam (Nijmegen) include stress measures (often based upon Guttman's rank-image disparities, d_{jk}^*), but also include a number of less familiar measures. In particular:

(a)   Mu $(\mu) = \dfrac{\sum d_{jk}d_{jk}^0}{\sqrt{\left\{\sum d_{jk}^2 \sum \left(d_{jk}^0\right)^2\right\}}}$

(Goodness of fit, varies between -1 and $+1$)

This measure is akin to the Pearsonian correlation coefficient. It is independent of the scale of both the distances *and* the disparities if the data are scaled by a ratio transformation, as in the metric MDS model (with the logarithmic option) implemented as MRSCAL in the MDS(X) series.

If the data are scaled by a linear transformation, as in MRSCAL (under the linear option), then mu is formally identical to the Pearsonian correlation coefficient r. It may also be used for monotone transformations, but it is not entirely clear whether it is dependent in this case on the scale of the disparities.

(b)   Alienation (K) $= \sqrt{(1 - \mu^2)}$ (Badness of fit, varies between 0 and 1)

This measure is akin to stress$_1$ and in some cases is identical to it. In any event, K is strictly monotonic with stress$_1$. The coefficient measures the extent of residual variance from the fitted regression.

(c)   (Normalised) Phi (ϕ) = (Raw Stress/$(2 \times$ NF 1))
$= \sum (d_{jk} - d_{jk}^0)/2 \sum d_{jk}^2$
(Badness of fit, varies between 0 and 1)

This measure is also akin to stress$_1$, but differs in the scaling factor—twice that of stress$_1$—and in the fact that the index is not reduced by its square root. It differs from the coefficient of alienation K, in being based upon the difference, rather than the scalar product, of the distances and the disparities.

Strictly speaking, any of these three measures may be used either with Kruskal's monotone regression disparities $\hat{d}_{jk}$, or Guttman's rank images d^*_{jk}, although by convention they are normally used with the latter.

4   Relation between fit measures

Relationships between the fit measures depend most importantly on whether Kruskal's $\hat{d}$ or Guttman's d^* quantities are being used. (In reporting measures of fit, users should always indicate which fitting quantities are being referred to and MDS(X) programs indicate the referent quantities as d-hat and d-star respectively.) In general, for any of these badness-of-fit measures, a measure based on d^* will be higher (indicating worse fit) than the same measure based on $\hat{d}$, since the former attempts to preserve strong monotonicity and the latter preserves only weak monotonicity with the data.

This can best be exemplified by relating various measures to μ, which represents the cosine of the angle separating the distances and the fitting quantities, d^0, in the measurement space (see Roskam 1969, p. 13):

	Fitting quantities (disparities)	
Measure	Kruskal's $\hat{d}$ (weak)	Guttman's d^* (strong)
μ	$\cos(d, \hat{d})$	$\cos(d, d^*)$
Stress$_1$ (S_1)	$\sqrt{(1 - \mu^2)}$	$\sqrt{2(1 - \mu)}$
Phi	$\frac{1}{2}(1 - \mu^2)$	$(1 - \mu)$
K (Alienation)	$(= S_1)$	$\sqrt{(1 - \mu^2)}$

Other useful relationships are as follows:

Alienation and phi	(i)	$K = \sqrt{1 - (1 - \phi)^2} = \sqrt{(1 - \mu^2)}$
Alienation and stress$_1$	(ii)	If $\hat{d}$ is used, $K = S_1$, and if d^* is used,
		$K = S_1\sqrt{1 - (\frac{1}{2}S_1)^2}$
Phi and stress$_1$	(iii)	$\phi = \frac{1}{2}S^2$

APPENDIX A3.2   CREATION OF THE INITIAL CONFIGURATION

A3.2.1   User-provided configuration

Most MDS programs give the user the option of providing an initial configuration. Usually, this will be a configuration thought to be close to the final configuration, either on a priori grounds or it will be a configuration from a similar study.

A3.2.2   Random or arbitrary start

The initial configuration may be a *random* start, formed simply by allocating random numbers to the $p \times r$ co-ordinates, or it may be an *arbitrary* start, positioning the points regularly at unit intervals along the dimensions of the initial configuration, as in the following 2-dimensional case, where they form a regular L-shaped configuration:

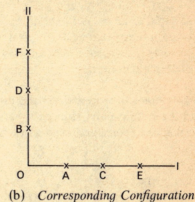

	Dimension	
Stimulus	I	II
A	1	0
B	0	1
C	2	0
D	0	2
E	3	0
F	0	3

(a)   *Co-ordinates*   (b)   *Corresponding Configuration*

In general, such a configuration can be produced from the series:

	Dimension				
Stimulus	I	II	III	...	r
A	1	0	0	...	0
B	0	1	0	...	0
C	0	0	1	...	0
⋮		⋮			
(r th)	0	0	0	...	1
(r + 1 th)	2	0	0	...	0
(r + 2)	0	2	0	...	0
⋮		⋮			
($2 \times r$ th)	0	0	0	...	2
	3	0	0	...	0
	0	3	0	...	0

... and so forth.

Such a configuration ensures that the co-ordinates are orthogonal, no matter what the dimensionality. Usually, the configuration is also centred and normed. It is the most common method used by Kruskal to create an initial configuration. It has the advantage that it in no way prejudices or influences the shape of the final configuration, but it is known to make the iterative procedure especially prone to sub-optimal (local minimum) solutions (Lingoes and Roskam 1973, p. 69) and should generally be avoided.

A3.2.3 Metric initial configuration

The data are treated as estimates of Euclidean distances, converted into scalar products, and the eigenvectors corresponding to the first r largest eigenvalues are used as the best (least squares) estimate of an r-dimensional initial configuration. This option is identical to Torgerson's (1958, pp. 254–9) classic method of metric MDS (see Appendix A5.2) and is closely allied to principal components analysis and Eckart-Young singular value decomposition. It generally produces a fairly good initial estimate of the solution, unless the configuration of points forms some highly non-linear shape (cf. Arabie and Boorman 1973). It is the strategy adopted by the programs TORSCA (Young 1968) and KYST (Kruskal et al. 1973) to form the starting configuration.

A3.2.4 Quasi non-metric initial configuration

Here the data are first reduced to rank order, thereby jettisoning all non-ordinal information. From these data a ranks matrix, $\mathbf{C}$, is formed:

$$c_{jk} = \begin{cases} 1 - \rho_{jk}/r & (j \neq k: \text{off-diagonal elements}) \\ 1 + \sum_l \rho_{il}/r & (j = k: \text{diagonal elements}) \end{cases}$$

where ρ_{jk} is the rank number of dissimilarity δ_{jk} and r is the maximum rank number.

The entries of $\mathbf{C}$ are similar to scalar products, and are a strict monotone function of data.

A principal components analysis is performed on $\mathbf{C}$, dropping the first (constant) eigenvector, $\mathbf{C} = \mathbf{FF}'$, and $\mathbf{F}$ is then the Guttman-Lingoes-Roskam initial configuration (see Roskam 1975, A7–8; Lingoes and Roskam 1973, pp. 17–19). Like the metric initial configuration, it will be quite close to the final configuration, and so will greatly reduce the number of iterations. Unlike the metric start, the quasi non-metric configuration has the advantage of using only ordinal information, and cannot therefore capitalise on possibly irrelevant quantitative (interval level) properties of the data.

In the MDS(X) series, MINISSA (and all programs in the Guttman-Lingoes-Roskam tradition) uses the quasi non-metric method of producing an initial configuration, and also allows users to input an initial configuration of their choice, if preferred. Although the quasi non-metric initial configuration certainly seems to guard best against suboptimal solutions, users are strongly advised to check solutions by using several different initial configurations.

4 | Interpreting Configurations

'Do you know what this is?'
'No' said Piglet
'It's an A'
'Oh', said Piglet
'Not O—A' said Eeyore severely

A. A. MILNE (The House at Pooh Corner)

The last chapter demonstrated how it is possible to construct a picture of a set of data in the form of a configuration of points, the distances between which represent the dissimilarity between the objects on which the data were collected. Such a representation is easier to assimilate than a matrix of coefficients, but we do not simply want to provide a picture but also, hopefully, to discover hitherto unremarked or unnoticed characteristics of the data. In other words we wish to interpret the configuration.

4.1 What Information is Significant and Stable?

Our interpretation is essentially a two-stage process. First, we look for significant patterns in the configuration, i.e. detect structure, and secondly we ascribe a meaning or interpretation to those patterns or structures. In so doing, it is important that the patterns make use of features of the configuration that are not simply arbitrary artefacts of the scaling procedure. The first questions, then, are: what information is significant and which aspects are stable in a configuration obtained from the basic MDS model? This restriction is important since output from other models sometimes preserves different significant information. The most basic significant information in the solution is the set of relative distances between the points. In the program these distances were calculated from a configuration which has a number of arbitrary, conventional, characteristics. In particular, the configuration was rotated to principal axes (see 3.5.5), and its actual size is also arbitrary, since it is the relative and not the absolute distances which are pertinent. This means that we may legitimately change the axes and the co-ordinate values so long as the relative distances remain unchanged.

The changes or transformations which may legitimately be made to a configuration in the case of the basic model are referred to as a similarity transformation; these consist of rotation, reflection, rescaling and translation (the three 'rs', with a foreign language for good measure?). These turn out to be of vital importance in Chapter 7 when we discuss the comparison of configurations. A detailed definition of similarity transformations is contained in Appendix A7.1 but an intuitive description can be given as follows.

Rotation of axes
MDS solutions using the Euclidean distance model are invariant under rotation—

93

i.e. the significant distance information is in no way changed by rigid rotation of the axes. *Any* set of axes which are orthogonal (at 90° to each other) will do as well as the axes given in the final configuration from an MDS program.

Reflection of axes
Consider a configuration reproduced on a photographic slide. It does not affect the distances in any way if the slide is looked at from the front or from the back, or indeed whether the slide is the right way up. Technically a reflection occurs when the positive or negative signs of the co-ordinate values on a given dimension are systematically changed.

Rescaling
The co-ordinates of the space may be *uniformly* rescaled, that is stretched or shrunk without destroying the significant information in the solution. This is simply another way of saying that relative distances are ratio-level quantities and may be multiplied by a constant.

Translation of origin
The origin, or zero point on all the axes, may be located at will within the space without changing the distances. (Note however that this does not apply for factor, or vector, solutions).

Clearly we do not want to interpret or make use of information in the configuration which may be affected by any of these transformations, and in the same way we do not want to make use of aspects of the configuration which are not stable.

We noted in Chapter 3 that the position of a particular point in the configuration was not uniquely fixed by the procedure, but rather fixed within a portion of the space known as its isotonic region, within which the point might be positioned without affecting stress. Obviously, the more data there are, the smaller such isotonic regions become. It remains the case, however, that whereas the overall (global) structure of a configuration is stable and reliable, local information (information about close or adjacent points) is not stable because of this freedom to move each point within its isotonic region.

We shall see that it is rarely possible to interpret the whole of a configuration. Rather, we will be discovering structure within parts of it. We want to be careful not to place too much stress (pun intended) on these local instabilities. Sometimes we can detect which points are most unstable in their location—either from studying the point contribution to stress (see 3.6.2.1), and/or by making several runs from different initial configurations and comparing the final configurations to see which points tend to change location. But in any event, analysis which relies upon small differences of location is not recommended, since it will almost certainly capitalise upon non-unique and unstable characteristics of the solution.

4.2   Internal and External Interpretation
Arrangements of points in a space do not normally exhibit any self-evident structure; we have to bring additional information to the task of interpretation. Two aspects will concern us particularly: pattern and meaning. Patterning refers to

the way in which points are located and related quite independently of what they may 'mean'. This may be evident by inspection or discovered by analysis. It is not usually difficult, for instance, to identify sets of points forming a straight line, or a circle, or a parabola, or even a set of discrete clusters—but it is more difficult to pick out general directions, or overlapping clumps, or even a 2-dimensional plane in a 3-dimensional space.

The 'meaning' of a configuration is a more complex matter. Once labels are attached to the points we bring all sorts of other information into play—what we know about the objects, what connotations they have for us, what subjects said about them and so on. As these meanings are put together we begin to recognise more subtle relationships. In fortunate circumstances, hitherto unsuspected characteristics of the data may then become apparent.

There is no procedure that will automatically detect structure in a configuration. The procedures described below will only assist the user to set about the task of identification in a fairly systematic manner, but there is no guarantee that the types of structure identified will be the most significant ones or relevant in any particular analysis.

It is worthwhile at this point to distinguish between *internal* and *external* methods of interpretation. In internal analysis only the original data are used in interpretation, whereas additional information is employed in the external case.

If we are to use only the original data in our interpretation we have two broad alternatives. Aspects of the original data may be represented in a graph-theoretic way, as line-segments within the configuration. This is a useful method by which simple structures may be identified. Alternatively, we may submit the same data to a clustering analysis and use this to interpret the scaling solution.

On the other hand, interpretation is often made easier by using information about the stimuli obtained independently of (externally to) the scaling itself. For instance:

in scaling subjects' judgments of similarity between nations one might use economic and political information on the nations concerned (cf. Wish 1972);

in scaling judgments of psychological stimuli (subjective loudness, brightness, colour saturation and hue), one would use information on the physical or objective variables involved (Carroll and Wish 1974);

in scaling judgments of personality-trait words, one might use known semantic properties and/or subjects' own ratings of the general properties of the words (Rosenberg and Sedlak 1972).

A closely related source of information for interpreting the meaning of a configuration, at least when the original data are obtained from human subjects, is what they themselves say when generating the original data. This is a much under-used resource, since a fair proportion of studies encourage subjects to verbalise as they produce data. Subjects' comments can be inspected and analysed in terms of their general semantic or particular substantive content and then related to features of the scaling representation.

4.3 Internal Methods of Interpretation

In this section we consider three aspects of configurations which have received particular attention: the *dimensions* (orthogonal axes which span the multidimensional space), the *regions* (or concentrations or high density of points, differentiated from others by empty regions) and the *simple structures* (identifiable one- and two-dimensional simple patterns).

4.3.1 Spanning dimensions

A set of orthogonal reference axes is necessary to locate any set of points. As we have seen, in the case of simple Euclidean distance the orientation of the axes is arbitrary yet a good deal of effort has been devoted, especially in the factor analysis tradition, to 'identifying' or naming them.

The important characteristics of a dimension, within this tradition, are that it represents a higher order organising construct ('factor'), which can be thought of as varying continuously and is bipolar (i.e. varies in both a positive and negative direction), and defines a major pattern of variation in the data (cf. Rummel 1970, ch. 21).

Having decided upon a set of reference axes, the dimensional analyst first separates out the objects or stimuli with the most extreme (positive and negative) co-ordinates on each dimension compared to those nearest zero, in order to establish which are the relevant and which the irrelevant objects for identifying the dimension concerned. Then the objects with the highest co-ordinates are compared to those with the lowest co-ordinates in order to identify the bipolarity of the dimension, or the contrast involved, if such there be.

The process of naming the dimension is by its very nature difficult to systematise. In effect, the researcher is performing a cognitive task analogous to that which social scientists often ask of their subjects (see 2.1). That is, 'In what way(s) do the high and low points differ?' Also, 'What property/properties do these points share, which others (in the zero position) do not possess?' The answer—the term used to label the dimension—depends in part on the researcher's verbal or conceptual abilities or on the accessibility of Roget's *Thesaurus*. In factor analysis and MDS, the most frequently encountered set of reference axes are principal components (see 3.5.5).

While it is useful to find out what the direction of maximum variation in the configuration actually is (the first principal component), there is no reason to suppose that it will be substantively significant or meaningful. On the other hand it may direct the user's attention to a readily identifiable, significant, general variable or factor underlying the configuration—general intelligence, general occupational prestige, the overall size of specimens, the left-right political continuum, the evaluative factor in word connotations, tough-tender-mindedness in personality studies—all have been much-heralded primary dimensions of variation.

What of the other dimensions of variation? It depends principally upon whether the user wishes to keep the dimensions orthogonal to each other. If so, a number of techniques exists for finding a set of reference axes which will often lead to a more interpretable set of dimensions, by (rigidly) rotating the principal components to a new position where, for instance, the co-ordinates of the points on a given axis tend to either unity or zero* (a 35° counter-clockwise rotation of axes in Figure 1.1

*Kaiser's 'varimax' rotation criterion. See Rummel 1970, pp. 391–3 or Maxwell 1977, p. 54 et seq. for a simple introduction and exposition.

above comes close to satisfying this requirement). If the MDS analyst wishes to pursue the identification of 'best' axes, then the detailed technology developed in the factor analytic tradition (cf. Rummel 1970, chs. 14–21) may be used. On any account, the naming of dimensions on purely internal criteria can be a hazardous and subjective undertaking, at least for the basic MDS model.

4.3.2 Graphical interpretation

A configuration may be interpretable in differing ways in different regions and there is no guarantee whatever that one particular type of structure will best describe the entire configuration. But if we hope to detect different types of local structure we shall need a procedure that is sufficiently general to cover other methods as special cases. A technique that has been found to be very serviceable in detecting different types of local structure is graphic analysis (Kendall 1971a; Waern 1971).

The basic idea is to represent the highest similarity values in the data by drawing a line in the MDS configuration between each pair of objects involved. This can be done either by deciding upon a cut-off value (say the top quartile of data, or all similarity values greater than 0.70) or by ordering the data from the highest to the lowest similarity values. If the data are ordered by size, the link between the highest similarity pair is drawn in the configuration first, then the next highest pair is linked and so on, until the researcher has sufficient information about the local structure. In this way the growth of various types of structure becomes apparent as the researcher moves down the ordered data list. These structures may include:

(i) *chains:* successive links which form a path through the configuration. The path may be (approximately) linear, signalling a vector or dimension-like property, or non-linear, a parabola, circle or some other simple regular pattern, or even an irregular, zigzagging, but connected, sequence.

(ii) *clusters:* links which occur within particular regions and build up to form locally-connected subgraphs or clusters based upon a few points.

In many instances both types of structure may appear, producing a linked set of clusters and often a residual set of isolated points.

4.3.2.1 Sequences and seriation

The historical scientist is naturally interested in inferring or detecting the time-sequence of a set of objects or events. This process is usually referred to as 'seriation' or 'ordination' (see Renfrew 1976, ch. 2; Hodson 1971, pp. 173–290; Hubert 1974). A good example of the success of the graphical method in finding a sequence is reported in Kendall's paper (1971b) in which he analysed an 'abundance matrix' consisting of co-occurrence counts of various artefacts such as type of pottery and jewellery common to pairs of graves in the neolithic cemetery of the la Tène culture at Müssingen-Rain. By using graphical analysis as described above, he showed that a chaining emerged whose shape led to its being described as a 'horseshoe'. Kendall hypothesised that the sequence of the points along the horseshoe indicated the historical order of the burials. This hypothesis was subsequently confirmed in an independent study by Hodson.

Why does a continuum (such as, here, time) become distorted in this way? There is no definite answer, but the evidence suggests that it is due in part to the fact that some permissible monotone transformations of the straight line generate a semi-circular MDS two-dimensional configuration (Shepard 1974, van de Geer, 1971, pp. 239–42). There is also increasing evidence (Kendall 1971a, p. 225) that it may be due to data collection procedures imposing an upper ceiling on values which the dissimilarities can take.

Thus, for example, in a rating exercise where only five or seven categories are allowed, the 'totally dissimilar' or 'totally unalike' category restricts the subject's ability to distinguish between very dissimilar and the very, very dissimilar. This, in turn, restricts the number of distinct values in the data matrix and the ability of the non-metric procedure to make the distance between points at opposite ends of the continuum as large as they 'should' be, thus producing the noticed horseshoe shape. This phenomenon can be overcome, i.e. the horseshoe sequence straightened out by treating the data as being only *locally* Euclidean, or by assuming that the largest distances are of least (or no) value in determining a solution, and relying only on proximate information. This may be done by using continuity mapping (5.2.2) or local monotonicity (5.2.1.1). The apparently simpler expedient of scaling in one dimension to recover such structures is usually misleading and is not recommended. This is partly due to the fact that highly irregular non-linear continua occur. As Shepard (1974, p. 386 et seq.) comments:

> In analyses of many different sets of data that were known to be basically one-dimensional, I have found that two-dimensional solutions, when attempted, characteristically can assume either the simple C-shape or the inflected S-shape ... Evidently, by bending away from a one-dimensional straight line, the configuration is able to take advantage of the extra degrees of freedom provided by additional dimensions to achieve a better fit to the random fluctuations in similarity data.

Users should therefore be on their guard against using one-dimensional solutions to recover an ordering and should use graphic procedures to help detect strong non-linearities of this sort.

Perhaps the best known, and independently replicated, instance of a C-shaped continuum in a 2-space is of adjudged musical intervals: the intervals separating the points correspond very well to the number of intervening semitones when they are projected onto the horseshoe, but no linear dimensional interpretation is possible (Levelt et al. 1966; Shepard 1974, pp. 386–7).

4.3.2.2   *Simple structures*

Other simple graphical structures have been studied and identified in scaling solutions, either from empirical patterns in the data or, in the Guttman tradition of facet theory (see 4.5) from the structure of the original mapping sentence (Lingoes and Borg 1979; Maimon et al. 1980). Among the more commonly encountered structures, in Guttman's terminology, are:

  the simplex (a chain, a linear or non-linear dimension or sequence);
  the circumplex (a circular arrangement of points);
  the radex (a combination of the simplex and the circumplex), consisting of two or
more concentric circles with lines emanating from the centre, dividing the circles

into sectors and thus the sheep from the goats. In three dimensions a set of stacked radexes is termed a *cylindrex*.

A simple example of facet-theoretic interpretation occurs in Levy and Guttman (1975) and a simplified version of the relevant mapping sentence is presented in section 4.5. The two-dimensional MDS solution for fifteen of the items on a USA sample of subjects is presented in Figure 4.1 and includes its radex interpretation. In such an interpretation the configuration is divided up into a number of regions within which particular object points possess a particular characteristic, or facet. In this instance, where the radex defines a number of sectors, item 15, life in general (LG), functions as the centre of the circle, the facet state *vs* resources form the inner and outer circular bands and the facet 'area of life' divides the circle into eight sectors corresponding to those of the original mapping sentence.

A similar simple structure was found in a re-analysis (Coxon 1974) of the Bollen-Delbeke data (Delbeke 1968) on preferences for families consisting of different size and composition (of boys and girls). Using the basic (distance) MDS model, the two constituent dimensions (number of boys and number of girls) were identifiable as two somewhat distorted lines, reflecting the empirical fact that people tend to prefer mixed family compositions. Using the vector model (see 5.3.2 below) for analysis of the same data, a radex structure was apparent. In this case, the inner and outer circular bands represented mixed *vs* unmixed composition, and the sector lines divided points in terms of the overall size of the family. These data and their analyses are discussed at greater length later, in sections 6.2.2 and 6.2.3.

A fascinating example of a cylindrex pattern ('stacked' circles) occurs in Heider

INTERRELATIONSHIPS AMONG FIFTEEN VARIABLES* OF SATISFACTION WITH LIFE AREAS IN THE UNITED STATES

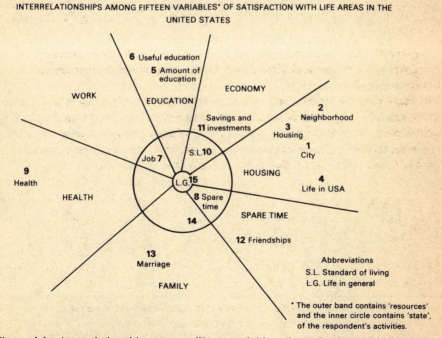

Figure 4.1   *Interrelationships among fifteen variables of satisfaction-with-life areas in the USA*

and Olivier's analysis (1972) of the structure of colour naming and memory in two markedly different languages. (The so-called Whorf hypothesis asserts that categories of naming used in different languages in some sense 'cause' different perceptions of reality, hence the interest of this study.) In this case, as in other studies of colour perception, the two basic dimensions of the space (brightness and hue) have in the MDS configuration become 'wrapped round' the brightness axis to form a circular pattern in 3-space, bringing together the red and purple ends of the hue dimension. The substantive interest in the Heider and Olivier study is to see whether the restricted categorisation of the Dani (a New Guinea people who use only two basic colour terms) affects their ability to recognise and identify different colour chips, compared to an English-speaking sample. Although the effects of restricted categorisation can be seen in the naming task configurations, it seems that retention of colour image in memory is unaffected by the very considerable cultural and semantic differences of the two languages. In both cases, colours of differing hue but the same brightness form a circular pattern, and the circles stack at different levels of brightness.

4.3.3 Regions of high and low density of points

Casual inspection and/or graphical analysis of an MDS configuration usually reveals that the points are not evenly distributed over the space. Rather, points tend to clump or cluster together, reflecting their high similarity, and are separated from other clusters by empty or sparsely-populated regions. That is, there exist subsets of points the relationships within which are stronger than those between them (Lingoes 1977, p. 116). Users may wish to check whether these clusters display any more formal structural properties.

A whole family of models, as extensive as those of MDS, exists for identifying and relating clusters of similar objects, namely cluster analysis (Wishart 1978; Everitt 1978). However, cluster analysis gives no information on the *extent* of separation of the clusters, and for this reason it is often advantageous to combine clustering analysis with MDS, which, of course, represents distances directly.* The best strategy is therefore to analyse the data *separately* by both a clustering and an MDS model and then represent the clusters *within* the MDS solution configuration. This is done in a 2-space by drawing a closed contour around the points contained in a cluster. The regions so enclosed will represent areas of high density, and the extent of their dissociation will be the distance in the configuration. This is done in a two-dimensional MDS plot: extension to three dimensions is usually not impossible, but poses problems in graphical portrayal.

Two varieties of clustering are extensively used in conjunction with MDS analysis—*hierarchical clustering schemes* (Johnson 1967, implemented as HICLUS in the MDS(X) series) and overlapping or *additive clustering* (Shepard and Arabie 1979, implemented as MAPCLUS; see section 8.2).

---

*Some data analysts recommend scaling followed by a clustering of the resulting *distances*. This practice is not be recommended since it capitalises on the weaknesses of both methods. As we have seen (4.1), MDS solutions are least stable in their fine-grain local structure, on account of the existence of isotonic regions, but this is the very information from which clustering initially proceeds and which significantly determines the final clustering.

4.3.3.1   *Hierarchical clustering schemes (HCS)*

An hierarchical clustering scheme takes a matrix of dissimilarity measures between a set of objects and represents the objects as being gathered into clusters on the basis of this information. It describes not one clustering but rather (for p points) p different clusterings, referred to as *levels* of a single total hierarchical *scheme*. At the highest level, all the objects are contained in one cluster, at the next highest there are two and so on until, at the lowest level, there are as many clusters as there are points. The defining characteristic of a hierarchical scheme is that once a point is incorporated into a cluster at a lower level it may not 'leave' that cluster at a higher one. Thus the clusters form a hierarchical scheme in the sense that each level is a special case of the next highest. We now consider in some detail the method of hierarchical clustering.

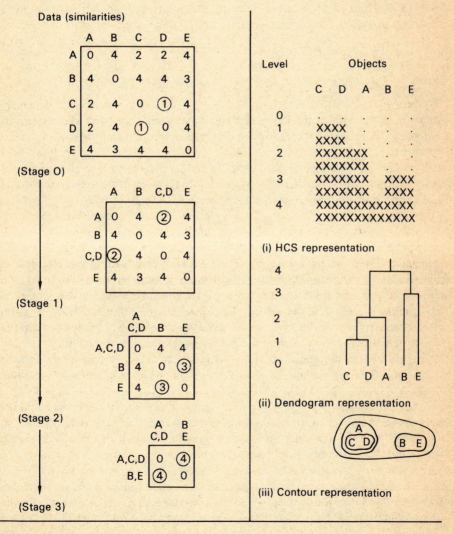

Figure 4.2   *Illustrative example of the HCS procedure and forms of representation*

The Method

Stage 0   The process of clustering begins by inspecting the original data matrix of dissimilarities and identifying first the most similar (or least dissimilar) pair of objects (C and D) and then merging them into a cluster. We now have the closest cluster of two points: (C, D).

Stage 1   Points C and D are from now on treated as a single *object* and the data matrix is reduced by removing the row and column of C and D and substituting one representing the cluster (C, D). In this example, the dissimilarity between C and each other object is the same as that between D and the same object—e.g. $\delta(C, A) = \delta(D, A) = 2$. (Normally this will not be the case.) The smallest entry in the reduced matrix, the currently most similar link, is now identified, and turns out to be between the cluster (C, D) and A. So the new structure consists of the dense cluster (A, C, D) and a set of unlinked points.

Stage 2   The new reduced matrix consists of the cluster (A, C, D) and points B and E. The smallest entry is now between B and E. This pair now form a new distinct cluster; the structure at this stage consists simply of the two clusters: (A, C, D) *vs* (B, E). A new reduced matrix is formed.

Stage 3   In this final stage, the remaining two entities—the two clusters (A, C, D) and (B, E)—are merged, forming the final clustering of all the points.

The process of hierarchical clustering—forming clusters at decreasing levels of compactness—gives considerable insight into the regions of the space. In this case, we can see that the basic contrast is between the (C, D, A) and (B, E) cores of clustering.

As in other areas of data analysis, especially block modelling of social networks (see White et al. 1976 and Breiger et al. 1975) 'holes', the empty areas, frequently turn out to be quite as significant as the clusters, and both aspects have to be represented in any structural analysis. Empty regions represent two types of significant information: differentiation or dissociation between clusters on the one hand, and/or the significant absence of objects on the other hand, which might mean that certain stimuli have been neglected or overlooked in a study or that no objects actually exist which have a particular combination of attributes.

Clearly, the most significant *clustering* information is contained in the initial stages, and the most significant *dissociation* information is contained in the later stages of a clustering.

In the above example there was no ambiguity in defining the distance between a newly-formed cluster and existing objects (clusters or points), but this will not usually be the case. Consider the simplest case where we have a cluster formed of two points A and B and a third point C. There will be two distances, namely those between A and C, $\delta(A, C)$, and between B and C, $\delta(B, C)$; and we have to decide how we are going to use these to define the distance between (A, B) and C, that is $\delta((A, B), C)$. If we want the procedure to produce identical clustering schemes when the data are monotonically transformed we cannot take the obvious step of averaging $\delta(A, C)$ and $\delta(B, C)$. Johnson (1967) suggests two contrasting ways of defining this distance in this instance:

The *maximum* method (otherwise known as the diameter or complete link method) defines the distance $\delta((A, B), C)$ to be the *maximum* of $\delta(A, C)$ and $\delta(B, C)$.

The alternative *minimum* method (also known as the connectedness or single-link method) conversely defines the distance between the new cluster and the extraneous point to be the *minimum* of the distances between the extraneous point and each of the points in the cluster.

When the data satisfy the ultra-metric inequality (see 6.1.6) and are therefore perfectly representable as an HCS, the two methods produce identical hierarchical clusterings. Otherwise, the two HCSs will differ—often not markedly, but sometimes significantly.

The maximum (diameter) method picks out the largest distance within a cluster as 'the' distance and seeks to minimise the diameter (largest distance between the objects) within a cluster. This tends to produce a fairly small number of compact clusters.

The minimum method, by contrast, selects the smallest distance as 'the' distance and seeks to minimise the largest link needed to produce a chain or connected path between the objects. It tends to produce rather a large number of broken clusters and is often marked by chaining—the continued addition of a single element to a cluster.

In practice, the minimum method is usually to be preferred to the maximum method in exploring the hierarchical structure of a set of data (although both methods should be inspected to see how far the data may legitimately be represented this way.* The chief use of the HCS procedure is to examine not only relatively dense 'local' structure of highly proximate points (the lower levels of the clustering) but also the open or 'global' structure of spaces which separate or dissociate the clusters (the highest levels).

Hierarchical clustering then, possesses a number of useful characteristics:

it presents not one, but a whole series of linked clusterings of increasing density, from a 'clustering' where each point is a separate cluster to the one where all the points are in a single cluster;

it includes two commonly used types of clustering as special cases and therefore gives the user some idea of how well the data fit the assumptions of the clustering model;

the HCS procedure is non-metric, in the sense that any ordinal rescaling of the data will produce identical results.

4.3.3.2 Clustering in high-dimensional space
The simple representation of HCS solutions within MDS configurations is only really feasible in two- or, at most, three-dimensional space. What if the user has a higher-dimensional solution and wants to gain some insight into the differential density of points in that space? An ingenious procedure is suggested by Andrews (1972; also see Everitt 1978, pp. 81–6) to represent each point as a wave form.

---

*Holman (1972) has shown that a set of errorless data will never perfectly satisfy both the Euclidean distance model and the hierarchical model, but will always satisfy one of the models to some extent.

Given a set of points in r-dimensional space, each point x is defined by its r coordinates: $\mathbf{x} = (x_1, x_2, x_3, \ldots, x_r)$. It is then represented as a Fourier series function of the form: $f_x(t) = x_1/\sqrt{2} + x_2 \sin t + x_3 \cos t + x_4 \sin 2t + x_5 \cos 2t + \ldots$, and the function is plotted for $-\pi < t < \pi$. The proximity of points can then be studied in terms of the similarity of the wave forms:

> If some plotted functions form a band by remaining close together for all values of t then the corresponding points are close together in the Euclidean metric.
> (Andrews 1972, p. 133)

Because the wave function preserves Euclidean distances the points which are close together have wave forms that have highly similar wave shape, whereas distant points have wave functions whose shape is different.

Because the wave function is affected by *all* the dimensions, the salient features of a high-dimensional configuration can be studied, and the procedure is therefore very useful for detecting isolates or outliers (which have markedly different wave forms) and clusterings (which have markedly similar wave forms).

4.4   External Methods of Interpretation

So far we have made use only of the original data in seeking to detect structure in the MDS solution.

When the researcher possesses additional external information about the points in the configuration, the task of interpretation is made much easier. These external variables, or 'properties' as they are often called, may come from a variety of sources. They may be relevant physical characteristics; they may be judgments made by respondents separately from the judgments used in the scaling; or they may simply represent the hunches or hypotheses of the researcher about the nature of the configuration. In any event, each property is assumed to consist of a set of numerical (interval or ordinal) values of the variable concerned, one for each point in the configuration.

A number of procedures exists for representing or 'embedding' each property within the already-obtained configuration, in a simple and easily recognisable manner. Two commonly used forms are as a *vector* and as a *point*. In essence a vector is a line drawn through the solution space and pointing the direction in which higher values of the property occur. Thus if our points were geographical sites, one relevant property might be 'northness', i.e. each point would have a value which was its north latitude. The property would then be fitted into the configuration map so that it was directed towards the pole. Such a representation would be adequate for points within areas of the northern hemisphere such as Scotland or the USA, but if the configuration actually contained the north pole, such as one consisting of sites in North America and Asia, in fact a map drawn from a vantage point above the pole, then the property of 'northness' would have to be represented not as a vector but as a *point* at the pole from which this property 'north' would decrease uniformly in all directions. Notice that representation in terms of vector or point depends on the characteristics not only of the property but also of the configuration.

These two forms of representing external information in a configuration are illustrated in Figure 4.3, with reference to the same 5-point configuration. Figure

4.3a represents a property of the points as a vector pointing in a north-easterly direction as the property values increase. Figure 4.3b represents another property in the same configuration of points, with the highest occurrences of the property somewhat left and below the origin of the space and systematically declining in all directions from this 'peak'.

These two ways of representing an external property within a configuration are now taken up in turn.

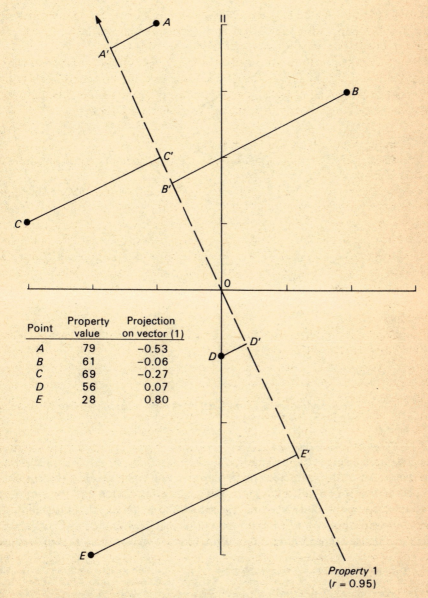

Point	Property value	Projection on vector (1)
A	79	-0.53
B	61	-0.06
C	69	-0.27
D	56	0.07
E	28	0.80

Property 1
(r = 0.95)

Figure 4.3a *Representation of an external property as a vector*

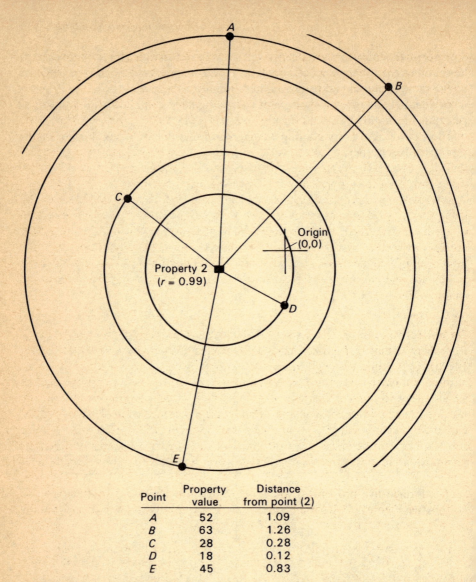

Point	Property value	Distance from point (2)
A	52	1.09
B	63	1.26
C	28	0.28
D	18	0.12
E	45	0.83

Figure 4.3b   *Representation of an external property as distances from a maximal point*

4.4.1 *Properties as vectors*

Locating a single property in a space as a vector is similar to finding a dimension in the space, except that a vector is unipolar and not bipolar as is a dimension. A property consists of a value for each point in a configuration and the aim of the procedure is to position the vector in the space (like an axis) so that the projections onto the vector (like co-ordinates on the axis) correlate with the property values in some well-defined sense.* Such a procedure has several uses: it accurately locates a

---

*Clearly, the match between the original property values and the projections on the property vector should be as close as possible. Current options in MDS(X) include maximising *ordinal* fit (PREFMAP IV with monotone option), *linear* fit (PREFMAP IV with linear option, or PROFIT with linear regression) and *non-linear 'continuity'* fit (PROFIT with continuity option). See section 6.2.1.

direction in the space for each property; it gives the researcher some assurance as to how accurate are her hunches about properties of the space, and it provides a useful way of mapping additional information into the space.

When more than one property is fitted into the same space, attention is focussed additionally upon the relationship between the fitted vectors. The basic information is quite simply the angle separating them. Thus a right angle represents independence (zero correlation), in which case they are equivalent to a set of a axes; 180° represents perfect *negative* association; and a 45° angle represents a linear correlation of 0.707 (cos 45°) and so forth.

Obviously, property-fitting can also be used to *identify* axes of a configuration by inspecting how close an external property comes to pointing in the same direction as the axis concerned. An example is presented in Carroll and Chang (1969, pp. 290–3), where they identify the three dimensions of a configuration obtained from scaling judgments of a set of tones as modulation frequency, modulation percentage and modulation waveform, to within 8°, 5° and 14° of the axes of the configuration.

An example

An instructive example of the uses to which vector property-fitting can be put is Rosenberg's studies of implicit personality theory (Rosenberg and Sedlak 1972a, 1972b). Each subject was given a set of 60 trait names (such as reserved, good-natured, submissive, humourless, etc.) and asked to sort them into groups, each of which was to represent a different person that they knew. A co-occurrence measure was used as a basis for constructing a dissimilarity measure between pairs of traits, and then scaled, yielding a two-dimensional solution, with stress$_1$ of 0.09. The subjects had also been asked to rate each of the 60 traits in terms of 5 general semantic differential scales (Osgood et al. 1965): hard/soft; good-intellectual/bad-intellectual; active/passive; good/bad; good-social/bad-social. The averaged ratings formed five properties, which were then fitted into the configuration as vectors (see Figure 4.4). Rosenberg and Sedlak's interpretation well exemplifies the use of property fitting and merits extended quotation:

> Five properties are obviously not needed to interpret a two-dimensional space. Moreover, there are alternative pairs of properties, all with high R (linear correlation) values, which can be used to interpret this space.
>
> It is possible, for example, to interpret the two-dimensional space in [the] figure [4.4] with the two general semantic differential factors, good-bad and hard-soft. Also, if we consider the three-dimensional solution, the R value for active-passive increases to 0.585 ($p < 0.001$), and the angle between the fitted axes for the three semantic differential factors are:

	good-bad	hard-soft
hard-soft	83°	
active-passive	92°	76°

Thus, the results from the three-dimensional solution support the presence of the three semantic differential factors in personality perception with each factor more-or-less orthogonal angles ($\cong 90°$) to the others.

An alternative interpretation, at least for the two-dimensional space in [the] figure [4.4] is the use of the two descriptive-evaluative properties, social good-bad and intellectual good-bad. It is interesting to relate this interpretation to Hays' (1958) findings that the extreme traits on one of his rank-order dimensions were warm and cold (social good-bad?), and on the second rank-order dimension they were intelligent and stupid (intellectual good-bad?).

These two descriptive-evaluative properties are not orthogonal, however, either in the two-dimensional (65°) or the three-dimensional (67°) spaces. Nevertheless, while orthogonality is convenient, it is not a necessary feature of an acceptable interpretation of a space. Nor are the two sets of interpretations of the two-dimensional solution incompatible. The heuristic distinction between connotative and denotative meaning is relevant here. The interpretation based on hard-soft and good-bad might be thought of as concerned with connotative meaning, whereas the interpretation of the space in terms of social and intellectual desirability is concerned with denotative meaning. A multidimensional analysis of the basic denotative meanings of traits may not result in orthogonal co-ordinates. Indeed, intellectual traits are likely to be perceived by college students as relevant for social activities.

(Rosenberg and Sedlak 1972b, p. 253)

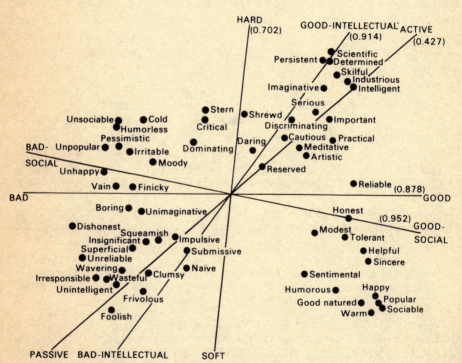

2 dimensional configuration of the 60 traits scaled by Rosenberg *et al.* showing the best-fitting axis for five properties. Each number in parentheses indicates the multiple correlation between projections on the best-fitting axis and property values. Reproduced by permission.

Figure 4.4   *Rosenburg's trait configuration interpreted by 5 property vectors*

When a property is represented as a vector the assumption is that the property is increasing across the space in the direction indicated by the vector and that the limit of the property is at infinity. Thus the lines joining points equidistant from the limit or ideal point are parallel to the vector. If the limit is brought in from infinity to the boundary of the space, i.e. the property is increasing but with a finite bound, then the equidistant lines assume the familiar convex shape of the indifference curves—iso-preference contours—of micro-economic analysis. If the limit or ideal point is brought within the boundary of the space, i.e. becomes not only finite but accessible, then the iso-preference or iso-similarity lines become circles around the point. As in our example of the property 'north' above, such a representation may make assumptions not only about the characteristics of the property but rather about the characteristics of the space. Thus, whereas in the case of vector representation the vector is positioned so that the perpendicular projections (co-ordinates) onto the vector of the points matches the property values, in the case of the point representations it is the distances from the 'property point' to each of the points in the configuration which are matched to the property values.

4.4.2 Properties as points

If the user collects data which relate a new object to each of the existing ones, and then interprets the information as distances, it is straightforward to locate it as a point in the original configuration (the SSAM and PREFMAP III programs allow just such an option). A useful application occurs when the user wants to position some new points in a configuration that is either already known or where the information for some points is more reliable than for others.

An example which illustrates such use is Tobler and Wineberg's study (1971), based upon the co-occurrence of the names of a number of Bronze Age merchant colonies in Cappadocia on a set of some 800 cuneiform tablets. The co-occurrence frequencies were taken to be a function of the size of the colonies and of their geographical separation, and the data were scaled in two dimensions. Unfortunately—but hardly surprisingly—the location of the great majority of the 65 Bronze Age towns finally used in the analysis were not known. But had the location of a significant fraction been known, they could have been treated as a known, fixed, geographical framework. The co-occurrence information for the remaining towns could then have been treated as a set of external properties and located as points within the known configuration. In this way it would be possible, in principle, to identify the location of colonies which had subsequently disappeared. (This is also illustrated in Kendall's famous paper (1971b), 'Maps from marriages').

Probably the most common use of property-fitting as points occurs in preference studies. In this case, a subject's numerical evaluations of a set of objects are located as a point of maximum preference in a configuration which has been previously obtained from judgments of similarity of the same objects. (This use is discussed in 5.3.3.1).

In all these instances—adding new objects to a configuration, mapping subjects' preferences as 'ideal points' or representing an external property as a point as an aid to interpreting a configuration—the basic principle is the same. The additional information is viewed as giving a set of relative distances from the new point to all the existing ones. The task of the scaling program is then to position the point so that it best reproduces those distances (or their rank order). To date, little use has been made of this way of representing a property as a 'high point' in the configuration, but it has considerable advantages.

4.5   Facet Theory as a Framework of Interpretation

There exists a methodology which views the whole of the measurement process, up to and including the interpretation of scaling solutions, as an integral whole. This is Guttman's 'facet theory', Space does not permit a detailed exposition here, and the interested reader is referred to Borg (1977). The core of facet theory is the 'mapping sentence' which relates various aspects ('facets') of the population of subjects, or objects/stimuli, and of alternative questions, and maps them into the actual response made. A simplified example relating to a quality of life study might be as follows (cf. Borg 1977, p. 75). (The bracketed and labelled components are referred to as facets, and within each facet the alternatives are separated by a slash (/) symbol.)

          A   *B*

The (*cognitive/affective*) assessment of the (*state of/government's treatment for*)

                                        C

the well-being of the respondent's social reference group (*self/government*)

                                       D

with respect to its (*primary internal/primary social/primary resource/ secondary state*) environment concerning a general aspect of life area:

                E

(*family/on the whole/economic/*. . .) according to the respondent's normative criterion for that life area

    IS MAPPED INTO THE RESPONSE-SET: (*very satisfactory/*. . .*/very unsatisfactory*)

An actual questionnaire item can be thought of as being an instance of a particular combination of these five facets (A, B, C, D, E). Consider for instance, the following three questionnaire items:

I   '*Generally speaking, are you happy these days?*' can be viewed as combining the facets $(a_2, b_1, c_1, d_1, e_2)$.* This specification in brackets is often referred to as a 'structuple'.

II   '*In general, how do you evaluate your family life?*' combines $(a_2, b_1, c_1, d_1, e_1)$.

III   '*What is your opinion of the way the government handles the economic problems of the country?*' combines $(a_1, b_2, c_2, d_4, e_3)$.

By inspecting the item specifications it is clear that items I and II are inherently most similar, differing only on the E facet (on the whole *vs* economic), whilst item III differs from items I and II and A, B, C, D as well. Clearly, facet theory makes the researcher be more specific about the content of her data collection procedures, and this is itself a good thing. It also makes it possible to inspect the *a priori* or theoretical similarity and relationship of questions *before* obtaining the data. The facet theory approach also formulates a number of useful expectations or

---

*Subscripts refer to the sequence number of the alternative within a facet.

hypotheses (or 'rules') about the data, e.g. 'The stimuli which are more similar in there facet structure will also be more similar empirically', which would lead us to expect their greater proximity in a distance model solution.

This abbreviated account is sufficient to show that the virtue of facet theory when it comes to interpretation of MDS solutions is that it alerts the researcher to characteristics to be looked for in the configuration, and to the type of structures (e.g. clusters of proximate items) to be expected. To this extent, facet theory can be used to assist the researcher to move beyond simple exploratory investigation towards a confirmatory approach.

In this chapter the simple structures to which Guttman and others have drawn attention have been discussed, but since they were originally developed within facet theory, interested users will profit from inspecting the full context (see Guttman 1971 and Lingoes 1977).

4.6 An Example of Interpretation: Occupational Similarities

An example drawn from one's own experience is most useful in conveying the detail and feel of the process of interpretation. I have therefore chosen work done in the Project on Occupational Cognition with Charles Jones, and concentrated on the interpretation of the basic configuration* of occupational similarities.

Data Pairwise judgments (using a 9-point rating scale) of the similarity of 16 occupational titles were obtained from 287 subjects. Both titles and subjects were selected from a fourfold typology of occupations chosen to contrast level of educational requirements and nature of the job—basically 'People' *vs* 'Data and Machines'.

Configuration Data were scaled in an aggregate (averaged) form, and in unaggregated form. The present example refers to the 2-D projection of the 3-D solution obtained from an Individual Differences Scaling (see Chapter 6) of a subset of 68 subjects' scalings. The basic configuration is that used as the basis for Figure 4.5.

The Resources available for interpreting the configuration were as follows:

    (a) *Researchers' 'internal' intuitions and hunches.*

    (b) *Data-based summaries for internal analysis*—in this case, the matrix of averaged similarities.†
These are presented in Table 4.1.

    (c) *Subjects' verbalisations.* As subjects completed their data ratings, they were encouraged to give the basis of their judgments, and these were either tape-recorded or written on the schedule. In each case, the comments were identifiable as referring to a particular pair of occupations.

    (d) *External information from subjects.* Subjects (including a number who had

---

*The substantive analysis is contained in Coxon and Jones 1978a, chapters 3 and 4 and the technical and methodological material is contained in chapters 2 to 4 of Coxon and Jones 1979. The data on which the analysis is based are available from the SSRC Survey Archive at the University of Essex.

†The original aggregate matrices (mean average, root mean square average and standard deviation of judgements) are presented in T3.8 in Coxon and Jones (1979).

provided similarities data) were also asked to rate and rank the same occupations in terms of a set of characteristics including:

1   Social usefulness.
2   Prestige and rewards they ought to receive
3   Social standing (involving the standard sociological definition in terms of general standing in the community)
4   Monthly earnings (estimated income), together with a measure of
5   Cognitive distance. This last characteristic is not considered further here.

The process of interpretation

(i)   *Dimension-naming*

The INDSCAL configuration is by definition already rotated to a non-arbitrary orientation (see 7.2.1), given by the east-west (Dimension I) and north-south

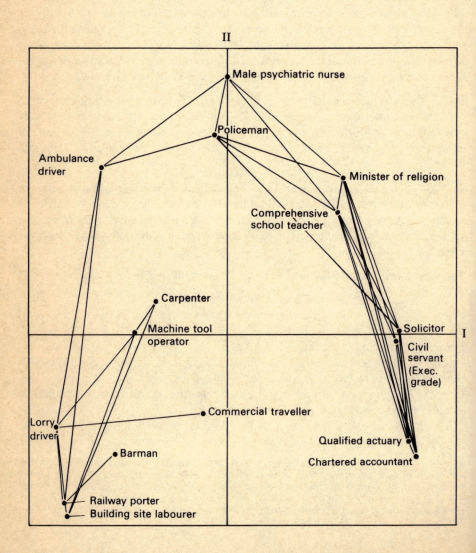

	MIN (1)	CST (2)	QA (3)	CA (4)	MPN (5)	AD (6)	BSL (7)	MTO (8)	SOL (9)	CSE (10)	CT (11)	PM (12)	C (13)	LD (14)	RP (15)	BM (16)
1 (MIN)	–															
2 CST	5.53	–														
3 QA	4.02	4.43	–													
4 CA	4.05	4.47	6.93	–												
5 MPN	4.44	4.53	2.92	2.65	–											
6 AD	3.20	2.76	2.09	2.01	5.13	–										
7 BSL	1.63	1.70	1.46	1.50	2.25	3.01	–									
8 MTO	1.93	2.57	2.11	2.16	2.71	3.25	4.23	–								
9 SOL	5.06	4.63	5.87	6.30	3.14	2.33	1.59	2.00	–							
10 CSE	4.43	4.89	6.10	6.23	3.09	2.35	1.47	2.05	5.71	–						
11 CT	2.91	3.10	2.98	3.28	2.81	3.88	2.50	2.86	3.03	3.06	–					
12 PM	4.27	4.27	2.81	2.90	4.67	5.17	2.47	2.78	4.18	3.69	3.26	–				
13 C	2.28	2.63	2.12	2.25	2.75	3.20	5.03	6.02	2.19	2.13	2.90	3.00	–			
14 LD	1.74	1.83	1.70	1.63	2.33	6.18	5.38	4.25	1.76	1.68	4.87	3.01	3.72	–		
15 RP	1.93	1.85	1.47	1.57	2.32	4.00	5.93	3.54	1.60	1.77	2.98	3.15	3.29	5.25	–	
16 BM	2.07	2.13	1.73	1.89	2.71	2.91	3.85	3.17	1.97	1.84	3.91	2.97	3.22	3.82	4.40	–

Table 4.1 Mean average similarities ratings

(Dimension II) directions on Figure 4.5. The occupational titles located at the extremes form the contrasts, and discontinuities are indicated by the gaps.

	Positive Pole	*vs*	Negative Pole	Discontinuity
		Contrast		
Dim. I	(CA, QA, CSE, CS)		(BSL, RP, LD)	MPN and CST
Dim. II	(MPN, PM, AD, MIN, CST)		(BSL, RP, QA, CA, BM)	C and CST

The terms used by the *subjects* to describe the contrast involved in Dimension I were retrieved by looking at what they said about the pairs concerned, e.g. Accountant *vs* Labourer; Accountant *vs* Porter; Civil Servant *vs* Porter; Solicitor *vs* Lorry Driver. The concepts they employed included 'qualifications', 'skills required', 'education', and we, as researchers, decided that the common core to the descriptions made the tag 'educational qualifications' a reasonable one. Of course, other connotations such as status and income were also present and other labels could just as easily have been chosen. Dimension II was named as 'service orientation' by a similar process. Note that, in the instance, we used subjects' descriptions of the contrast as a fundamental resource, but decided upon the final identifying label ourselves.

(ii)   *External properties*
In the study, we treated occupational judgment as involving the analytically separable components of cognition and evaluation, the former operationalised in terms of similarity judgments and the latter in terms of the first three characteristics listed earlier. The subjects' ratings were then averaged, thus providing four 'external properties'.

Both point-distance location and vector representations were used (by means of the PREFMAP III and IV models). Without any doubt the vector representation gave a better fit to the data for each property, and these are presented in Figure 4.6a. An interesting point is that the most explicitly evaluative property, 'social usefulness', is independent of (at right angles to) estimated earnings: on average, people judge the pay for an occupation to be unassociated with its social worth. Moreover, the 'good' end of the vectors—whether of social utility, status or earnings—all point towards the professional side of the configuration.

Turning now to the differential regional density of the points, it is worth noting that, if the largest single gap on each spanning dimension is marked as a dividing line, the four quadrants each include the two occupations chosen in the original design to be examples of a four-fold typology combining 'educational require-ments' and 'people-orientation'. (These occupations are emphasized in Figure 4.6b.) In effect, the subjects view the occupations very similarly to the researchers, which is a useful bonus.

The diameter HCS clustering of the averaged similarities data is presented in contour form in Figure 4.6c, up to the final two levels. The pattern is clear with either HCS: the major divide is between the left and right hand sides the manual and the professional occupations, with the Barman and Commercial Traveller largely unassociated.

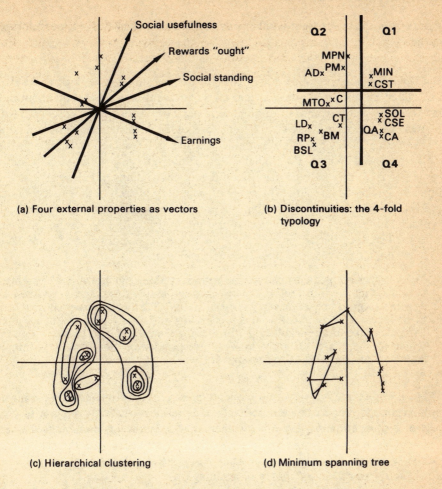

(a) Four external properties as vectors

(b) Discontinuities: the 4-fold typology

(c) Hierarchical clustering

(d) Minimum spanning tree

Figure 4.6 *Various interpretations of single configuration*

The dense clusterings occur round the two professional groups (people- and non-people oriented) and within the skilled and unskilled manual groups. It is very obvious that the clustering follows very closely the horseshoe sequence shown in Figure 4.5 and not any dimensional direction.

(iii) *Graphic analysis*

Finally, graphic analysis was used to detect local structure and it was at this point that the horseshoe pattern became very evident. When mapping the lines into the configuration in sequence from the highest similarity down, it became very obvious that the pattern was that of a linked set of clusters, rather than a simple sequence of occupations. In addition to mapping the top quartile of similarities, the minimum spanning tree (MST) was also constructed. The purpose of constructing a MST (Prim 1957) is to connect the points by a network (or tree) of *connected* links which have the smallest *overall* distance. The MST is illustrated in Figure 4.6d. Once again, the horseshoe sequence is very apparent as the first ten links join up the points from the lower right-hand corner *in sequence* round to the lower left-hand corner.

(iv)   *Final interpretation*

But what does the horseshoe mean? The clue to this came from looking at the HCS and matching clusters of points at each level with what the subjects said about each cluster. Thus Actuary and Accountant were regularly described as 'concerned with figures'; when Solicitor and Civil Servant are added as 'instrumental occupations' or 'managing', and when joined with the Minister and Teacher ('teaching') and the Policeman and Nurse ('custodial'), the entire group are repeatedly called 'professions'. It thus became clear that different descriptions (predicates) existed at different levels of generality. The range of generality of each of the most common predicates was obtained by finding out to which occupations a given predicate typically applied. This was looked at and mapped into the horseshoe sequence, with the length of the arrow indicating the range of generality (see Figure 3.6 in Coxon and Jones 1978a). It then became clear that:

> The predicates repeatedly change as one moves along the 'horseshoe', making it very difficult to interpret the map as only involving a single contrast or dimension, even one as general as 'social status'. Yet there is a sequence, at least in the sense that predicates tend to develop and rarely appear outside their 'typical' range of applicability ... In moving from cluster to cluster along the sequence there are certainly many correspondences but equally, common features drop out and others re-appear.

> (ibid, p. 92)

Thus, while there is a very clear sequence present in the configuration—a highly non-linear one, not to be confused with a dimension—it turns out not to be a sequence of continuous meaning. Rather, it is a family of resemblances where characteristics or properties apply at different segments so that the sequence is more like a chain of associations or what Wittgenstein (1958, section 60) describes as a 'family resemblances' theory of meaning.

However abbreviated, this example makes it clear that different methods of interpretation often provide convergent evidence about both local and global aspects of the structure or configurations which, with a little imagination, can be invaluable in aiding interpretation.

5 | Characteristics of MDS Models

> *The old order changeth, yielding place to new*
> *And God fulfils himself in many ways,*
> *Lest one good custom should corrupt the world*
>
> TENNYSON (Idylls of the King)

We have used three basic characteristics to define the basic MDS model. These are:

Type of data
Transformation (rescaling) function (or level of measurement)
Form of model.

In this chapter we shall use these characteristics to differentiate and describe other programs in the MDS(X) series. Programs from other sources can be described in precisely the same way. This three-fold characterisation is akin to the new typology of scaling programs developed by Carroll and Arabie (1980) in their definitive review of scaling developments of the past seven years.

We begin by giving a fully-specified description of the basic model and then proceed to a discussion of the various ways in which types of data, transformations and models can be extended. These characteristics will be used to describe the MDS(X) programs for the analysis of 2-way data in the next chapter and for the analysis of 3-way data in Chapter 7.

The basic model: a fuller specification
The basic model has already been defined (section 3.1) as follows:

BASIC NON-METRIC MODEL
The analysis of

(Characteristic)	(Specification)
(DATA)	A square symmetric 2-way table of (dis)similarities
(TRANSFORMATION)	by a monotonic rescaling function
(MODEL)	using a simple Euclidean distance model

Enlarging on this specification we might describe the basic model as in Table 5.1. Each of the aspects appearing here is taken up in the following section.

5.1 Data
By data in this context we simply mean information input to the program. Even for

BASIC NON-METRIC MODEL		EXTENSIONS
Characteristic	Specification	
DATA	The INTERNAL analysis of a 2-WAY SQUARE (Unconditional) SYMMETRIC matrix of (dis)similarities	External 3- and higher way/modes Non-square, rectangular, triadic ...
TRANSFORMATION FUNCTION	By a GLOBALLY MONOTONIC rescaling	Locally Linear, Power, Continuity ...
MODEL	Using a SIMPLE EUCLIDEAN DISTANCE model	Weighted Other Minkowski and non-Minkowski metrics Additive, Subtractive, Multiplicative (scalar product/factor) ...

Table 5.1   *Fuller specification of basic MDS Model and extensions*

the basic model, which assumes a square, symmetric (or lower triangular) array of data, the information might either be pairwise similarity ratings obtained directly from subjects, or else indirect measures of co-occurrence, covariance, contingency, association etc., obtained by aggregating over simpler data, as we saw in Chapter 2. The *source* of the data does not concern us; the *form* of the data and its interpretation do.

The distinctions referring to data which are made in Table 5.1 are between:

(i)   the way and mode of the data; and
(ii)   internal and external analysis

We shall consider each in turn.

5.1.1   The 'way' and 'mode' of data

The 'way' of data is simply the dimensionality of the data array. 'One-way' data would simply consist of measures on a single set of objects, such as *one* individual's set of preference judgments of the loudness of a set of tones, or the frequency of a particular plant species on a set of geographical sites. Since one-way data are never scaled as they stand, they need not be examined in detail here.

Two-way data take the form of a matrix consisting of rows and columns, and relate a pair of entities. To say that a set of data is two-way says only that it may be represented in a single matrix. It does not tell us whether the matrix is square or rectangular, symmetric or asymmetric.

In order to make such distinction, the notion of mode is introduced. In a two-way matrix, the rows and columns may refer to the same set of objects or to distinct sets. If the rows and columns refer to the same entities—and the matrix is thus necessarily square—then the matrix is said to have *one* mode, the one set of entities represented. If, on the other hand, rows and columns refer to two distinct sets, then the data are said to have two modes. The mode of the data therefore is the number of distinct sets of entities to which it refers.

Normally, of course, two-way, two-mode data form a rectangular matrix with, conventionally, 'data producers' (individuals, groups, locations) as row-elements, and objects (stimuli, attitude statements, symbolic entities) as column-elements; but two instances where this is not the case are apparent, which clarify the usefulness of this way/mode distinction. The first is where the number of row-elements happens to be equal to that of the column-elements, and the second where the row- and column-entities are in fact the same but are considered distinct for the purpose of analysis, e.g. firms considered as producers and consumers, members of a group as rankers and ranked in a sociometric exercise.

The extension to higher ways and modes should be obvious, and these are considered later in Chapter 7.

5.1.1.1 Asymmetric data

Of particular interest are those data matrices where the row and column elements happen to refer to the same objects—so the matrix is square—but where the elements δ_{jk} and δ_{kj} are considered distinct. Such data occur as sociometric rankings, occupational-mobility turnover tables, economic input-output tables, migration and communication flows, citations within and between journals, and confusion between pairs of auditory stimuli presented in a left-right and right-left order.

At first sight it may seem perverse to wish to represent such data by what is, after all, by definition a *symmetric* distance model. Several ways have been proposed to deal with this anomaly:

(i) to treat the asymmetry of δ_{jk} and δ_{kj} as 'noise' or chance error and simply symmetrise the data by replacing the corresponding entries by their median, or by the arithmetic or geometric mean. Such a treatment, of course, simply defines the problem out of existence and the resulting symmetrised data matrix can now be analysed by the basic model.

(ii) to treat the asymmetric information as consisting of two distinct components, each of which is capable of being represented separately by the distance model: the 'flow' from j to k and the 'flow' from k to j. This alternative was discussed in section 2.2.3.4 where the index of dissimilarity was used to compare both row percentages (outflow) and column percentages (inflow). Typically, the two resulting matrices of outflow and inflow coefficients are scaled separately by the basic model, and the solutions are then compared (see, for instance, Macdonald 1972, pp. 214–27 and Blau and Duncan 1967, pp. 67–75).

(iii) to treat the asymmetry as arising from the conditional nature of the data, but not as a characteristic needing separate representation. The entries within the same row of the matrix will be treated as being comparable, but information between rows will not.

This interpretation is most relevant where data have been collected by the method of conditional rank orders (see Rao and Katz 1971, p. 470) or, in Coombs' terminology, 'order $(p - 1)$ out of p stimuli'. In this method the subject is presented with each stimulus in turn. She is then asked to rank each of the other stimuli in terms of their similarity to the reference stimulus, thus generating what amounts to a set of p I-scales, with each stimulus in turn serving as the 'ideal point'.

In this last instance, the stimuli are represented as a *single* set of points, although the entries δ_{jk} and δ_{kj} will normally be fit by distinct disparity values. (This is the model fit by MINICPA described in section 6.1.2.) Another alternative is to treat the row and column elements as distinct points. Thus each 'stimulus as subject' (rows) and each 'stimulus as object' (columns) will be represented as separate points. This option also treats the data as providing conditional distance information and is identical to the unfolding model described in 5.3.3.1. It is implemented by the MINIRSA program.

(iv)   to treat the asymmetry as something extrinsic to the distance information and represent it in some other way. A number of ways have been proposed, including representing asymmetry as contours and as 'jet-stream' directions over a conventional scaling configuration, and are discussed in Gower (1977).

(v)   to interpret the data as a graph with each distance represented as a link between two points, allowing the distance $i \rightarrow j$ to be different in length from $j \rightarrow i$ (see 6.1.2).

5.1.2   *Internal* vs *external analysis*

The distinction between internal and external analysis was made in Chapter 4 with regard to the interpretation of configurations. There we noted that in internal analysis it was the original data only that were used in the interpretation while additional information was brought to bear in the external case. Generally speaking, internal analysis, or 'unconstrained' solutions (Carroll and Arabie 1980), uses only the information given to generate the solution, while external analysis ('constrained' solutions) takes one part of the input as fixed and relates the data to that fixed 'external' part.

5.2   Transformations

Whilst the full range of Stevens' levels of measurement may, in principle, be used in scaling, only a small number have in fact been used, and in the MDS context the only ones which concern us directly are the nominal, ordinal, interval and ratio levels. By and large, most data are at the nominal and ordinal level—or researchers with justifiable caution consider their data so to be—whereas most scaling solutions are at the ratio level (e.g. distances) or occasionally at the interval level (e.g. solution scales from conjoint measurement).

The transformation function, rescaling the data into distances, normally matches the level of measurement of the data. For our purposes, the most important transformations are the monotonic (ordinal) and regular (linear or logarithmic) rescaling functions, but we shall also consider the 'continuity' or 'smoothness' transformation, which does not fit easily into the conventional levels of measurement, having affinities with both monotonic and metric scaling.

Although nominal rescaling functions have been developed† they are not widely employed in programs in the MDS(X) series other than ssa(m) and are not considered further.

5.2.1 Monotonic transformations
The monotone relation is best illustrated by the Shepard diagram (Figure 3.11 et seq.), where fitting values (d^0) are joined up to form a jagged monotonic 'curve'. The line segments drawn to join up the fitting value points are simply an aid to visualising the relationship and show that the relationship between the data and distances is ascending (in the case of dissimilarities) or descending (in the case of similarities). But the slope of the segments has no intrinsic meaning whatever since it depends in part on the purely arbitrary ordinal scale of the data.

In most MDS applications the shape of the monotone curve is characteristically very jagged and 'steppy', especially if there are ties in the data and weak monotonicity has been used as a criterion in monotone regression (see section 3.2.3). But users should be alert to signs of smoother regularity in the monotone curve. Two particularly important, and more regular, forms of the monotonic function are the straight line and power function curves (including the exponential and logistic curves). All of these variants of monotone relationships are illustrated in Figure 5.1.

An actual example of a monotone regression function approximating a linear function is seen in the Shepard diagram of Figure 3.2; the co-occurrence data scaled in section 3.6 provide a fair approximation to a (negative) exponential function (Figure 3.14b), and the relation between the rank of a mileage and its recovered distance exemplified in the Scottish mileages data is very well approximated by a logistic function (Figure 3.4). Whenever a more regular function is discerned, the data should then be re-analysed using the appropriate scaling transformation.‡ (Linear and power transformations are permitted in the mrscal program).

5.2.1.1 Local and global monotonicity
The monotonicity criterion requires that *all* the data should be monotonic with the distances (global monotonicity). On occasion this may be thought too restrictive and monotonicity be only required locally, i.e. around the neighbourhood of each point, hence the term 'local monotonicity'.

A familiar example of the local monotonicity principle occurs in geographic mapping where 'stereographic projections', which seek to represent the earth's three-dimensional surface as a two-dimensional map, are commonly used. Clearly, this cannot be done without some distortion, and the various geographic projections preserve different aspects of distance. The 'conformal mapping' projection involves a principle very similar to local monotonicity, since smaller distances are accurately represented but larger ones are not. Hence the distortion is

†Nominal rescaling functions are employed in the Multiple Scalogram Analysis option in ssa(m) and in allied programs in the Guttman-Lingoes series (Lingoes et al. 1979, pp. 274–7; Zvulun 1978) and nominal rescaling is permitted as an option in the alscal program (Takane et al. 1977) for analysing 2- and 3-way data.

‡Shepard (1974, p. 395 et seq.) describes a number of other approaches to constraining the monotone function to convexity, concavity, smoothness, etc.

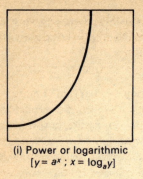

(i) Power or logarithmic
$[y = a^x ; x = \log_a y]$

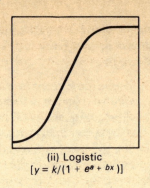

(ii) Logistic
$[y = k/(1 + e^{a + bx})]$

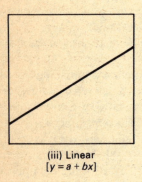

(iii) Linear
$[y = a + bx]$

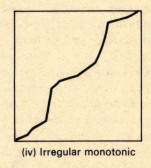

(iv) Irregular monotonic

Figure 5.1   *Monotone curves*

concentrated in the largest distances and this fact needs to be taken into account in reading the map. The same principle applies in MDS solutions derived by use of this criterion.

There are two cases where it is useful to use the local monotonicity criterion, one being when ceiling effects occur in the data producing the familiar C-shape, or horseshoe shape, discussed in section 4.3.2.1. To overcome this effect, the remedy is fairly simple: ignore or down-grade the importance of the largest data dissimilarities. This can be implemented as follows:

(i)   by choosing the local monotonicity option in SSA(M); or

(ii)   by using a program such as PARAMAP which implements 'continuity' transformation, one of whose features is to act like a local monotonicity constraint (see 5.2.2 below).

Both options have similar and often dramatic effects—in 'unbending' highly non-linear simple structures.

The second use of local monotonicity is to map a high-dimensional solution into a space of lower dimensionality. This procedure is acceptable if local structure is of primary interest and larger distances which will be distorted can be ignored. It should be noted, however, that Graef and Spence (1979) have shown that the largest distances do most work in producing an MDS solution and they can be critical in the satisfactory recovery of a configuration.

5.2.2 Continuity (smoothness) transformations

In basic non-metric scaling a best overall monotonic fit is sometimes achieved by producing sudden changes in distance values which do not exist in the data values. If we are firmly committed to the assumption that there really is no information in our data other than the order of the dissimilarities, well and good. But if we believe that the data contain more than simple ordinal information then these sudden discontinuous jumps may well distort the local structure, producing high distance values to correspond to very close data values. In this case, it might be better to concentrate on minimising or smoothing out the jumps by making the relationship between the data and the distances of the solution as 'smooth' or 'continuous' as possible, even at the cost of worsening the overall monotonic fit. This can be done by requiring that when two data values are close to each other, then there should be little difference (or variation) in the corresponding distance.

This basic idea of continuity can best be illustrated by a simple example of a one-dimensional solution. Suppose we wish to examine the relationship between the physical loudness of a set of six tones, x_1 to x_6, and their perceived loudness, y_1 to y_6. Our attention will concentrate, as we move up the scale, upon whether perceived differences in loudness change in the same manner as physical differences do.

> If we say that the y values seem to change in a 'continuous' manner as we move along the underlying x continuum, we are essentially saying that the change in y as we move from one x value to the next tends to be small compared to the change in y generally associated with larger jumps in x.
>
> (Shepard and Carroll 1966, pp. 579–80)

In Figure 5.2 two examples of such a relationship are given: one where small changes in x are *not* associated with small changes in y, a relatively discontinuous relationship (Figure 5.2a), and another (Figure 5.2b) where small changes in x are associated with small changes in y, a relatively continuous relationship.

The extent to which the relationship between x and y is smooth or continuous can be monitored by a simple index which compares changes in y to changes in x, for each of the adjacent pairs along the scale. This can be done by taking the differences between adjacent values of both x and y and finding the ratio:

$$\frac{\text{Difference in } y}{\text{Difference in } x} = \frac{\Delta y}{\Delta x} = \frac{y_j - y_k}{x_j - x_k}$$

Clearly, when the relationship is smooth or continuous, Δy and Δx will be almost the same and the ratio will be about 1. But if small changes in x produce large changes in y, then the ratio will be correspondingly large. A simple overall measure of discontinuity (DISCONT) is constructed by squaring the ratio (both for computational simplicity and to draw attention to particular gross discontinuities) and then summing over the adjacent pairs:

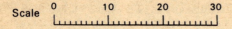

Scale

Adjacent pair	Ratio of diffs:* $\Delta y / \Delta x$	Squared ratio
a b	16 / 2	64.00
b c	2 / 11	0.03
c d	8 / 4	4.00
d e	2 / 9	0.05
e f	2 / 4	0.25

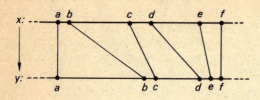

DISCONT (Sum) = 68.33

(a) Discontinuous

Adjacent pair	Ratio of diffs:* $\Delta y / \Delta x$	Squared ratio
a b	3 / 2	2.25
b c	11 / 11	1.00
c d	6 / 4	2.25
d e	7 / 9	0.61
e f	3 / 4	0.56

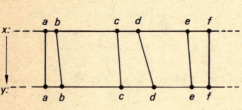

DISCONT (Sum) = 6.67

(b) Continuous

*i.e. $(y_i - y_j) / (x_i - x_j)$

Figure 5.2   *Discontinuous and continuous relations between two continua*

$$\text{DISCONT} = \sum_{\substack{j,k \\ \text{adjacent}}} \left(\frac{\Delta y}{\Delta x}\right)^2 = \sum \left(\frac{y_j - y_k}{x_j - x_k}\right)^2$$

This measure is calculated in the boxes alongside the two examples in Figure 5.2. The relatively continuous relation (B) has a value close to 5 (the minimum value of DISCONT), and the discontinuous relation (A) has a value of 68.33. (Note that in this latter case the value of DISCONT is most affected when small differences in x are accompanied by large differences in y, e.g. for (a, b) and (c, d). By contrast, where large changes in x give rise to small changes in y the contribution is very small, and this measure will be largely insensitive to them.)

5.2.2.1   Kappa as an index of continuity

The continuity transformation is used in MDS to obtain a solution where differences in the data correspond as smoothly as possible to differences in the solution differences. To do this we need the use of measure such as DISCONT, rather than stress. But in adapting DISCONT to measure the discontinuity between *multidimensional* spaces (rather than uni-dimensional continua) we run into a problem. With a single line, the idea of a small change in value as we move up the continuum is easily defined: it is the difference between *adjacent* object locations. The notion of 'difference' generalises perfectly easily to 'distance' in the multidimensional case, but a little thought will convince you that there is no equivalent to 'adjacent' points in a two (and higher) dimensional space. But there is an approximation that will suffice: 'adjacency' can be replaced by 'closeness' or 'relative proximity' so long as we take care that only information relating to the immediate vicinity of each point is taken into account. In constructing an index of discontinuity in the multidimensional case, we shall therefore want to emphasise the distance involving closely proximate points and successively de-emphasise those at increasing distance. (This is obviously a further instance of *local monotonicity* described in the previous section.) In the context of MDS, the DISCONT measure is known as the 'kappa' index, symbolized by κ. The simplest measure on the analogy of stress, is referred to as 'raw kappa' and consists of two components, a discontinuity ratio and a weighting factor which restricts attention to the most proximate points:

$$(\text{raw}) \text{ kappa} = \begin{array}{c} \text{discontinuity} \\ \text{ratio} \end{array} \times \begin{array}{c} \text{local proximity} \\ \text{weighting factor} \end{array}$$

$$\kappa = \left(\frac{\Delta x_{jk}}{\Delta y_{jk}}\right)^2 \qquad \times w_{jk}$$

Discontinuity ratio

In MDS applications we wish to ensure that small changes in the solution distances (d_{jk}) are associated with small changes in the data (δ_{jk}). Working with squared distances, as in DISCONT, the ratio becomes*:

$$\sum_{j \neq k} \sum (\delta_{jk}^2 / d_{jk}^2)$$

Weighting factor

In the case of kappa, the weight factor is made the reciprocal of the corresponding squared solution distance:

$$w_{jk} = 1/d_{jk}^2$$

This form of weight has two useful properties: it ensures that local monotonicity is preserved (decreasing the contribution of any pair by the square of its distance, so proximate pairs contribute a good deal, and far distant ones scarcely at all), and the weights remain invariant under changes of scale.

Put together, these form the raw kappa index:

$$\text{Raw } \kappa = \sum_{j \neq k} \sum \left(\frac{\delta_{jk}^2}{d_{jk}^2}\right)\left(\frac{1}{d_{jk}^2}\right)$$

or, in simplified form:

$$\boxed{\text{Raw } \kappa = \sum_{j \neq k} \sum \left(\frac{\delta_{jk}^2}{d_{jk}^4}\right)}$$

*See Shepard and Carroll 1966, p. 581 et seq. In their treatment, data are referred to as (d_{jk}^2) and solution distances as (D_{jk}^2).

As in the case of raw stress, this index has the unfortunate property that an arbitrary enlargement of the solution configuration can make departure from continuity (raw kappa) as small as desired. And once again, the remedy is a normalising factor that will ensure that changes in the scale of the solution do not affect the index. Shepard and Carroll (1966, p. 583) show that the simplest effective normalising factor is:

$$\text{NF} = 1 \bigg/ \left(\sum\sum_{j \neq k} \frac{1}{d_{jk}^2} \right)^2$$

The normalised index of discontinuity (used in PARAMAP and non-linear PROFIT) then becomes:

$$\text{Normalised } \kappa = (\text{Raw } \kappa)/\text{NF}$$

$$\kappa = \sum\sum_{j \neq k} \frac{\delta_{jk}^2}{d_{jk}^4} \bigg/ \left[\sum\sum_{j \neq k} \frac{1}{d_{jk}^2} \right]^2$$

These, and related measures, are further discussed in Appendix A5.1 and in the PROFIT and PARAMAP documentation of the MDS(X) series.

By minimising kappa, continuity scaling both preserves local structure and allows solutions to be forced down into very small dimensionality, so long as the user is prepared to disregard or downgrade large distances. The Shepard diagram resulting from continuity scaling has a characteristic fan-like form which reflects these properties. As the (solution) distances increase, the corresponding data values increase, which reflects the fact that any discrepancy in the representation of small distances is heavily penalised (i.e. local structure is being preserved), whereas even very large discrepancies in representing the largest distances are virtually ignored. Typical examples of the diagram occur in Shepard and Carroll (1966, p. 575) and in Coxon and Jones (1978b, p. 266), reproduced as Figure 5.3.

Continuity scaling is a hybrid transformation. In that it assumes that the data are a direct estimate of the solution distances (except for a possible scaling factor), so it implicitly assumes that the data are at the ratio level of measurement, and is therefore an instance of classic metric scaling (see section 5.2.3.2). But it also preserves local monotonicity, and to that extent continuity scaling can be viewed as an even weaker form of monotonicity than that assumed by non-metric scaling. However, the continuity criterion ensures that the characteristic 'steepiness' and 'angularity' of the monotone function are smoothed out.

5.2.3 Regular transformations

By 'regular' transformations we mean those which are expressible in a simple mathematical form and are systematically increasing or decreasing. In effect, the term covers ratio and linear—often confusingly called 'metric'—and power rescaling functions. Regular rescaling transformations have the advantage over irregular monotonic transformations of being smooth and simple in form. Hence if the researcher's main interest focusses upon the relationship between the data and the underlying model, rather than on the solution itself (as, for example, in studying the relation between physical and perceived properties of colour or

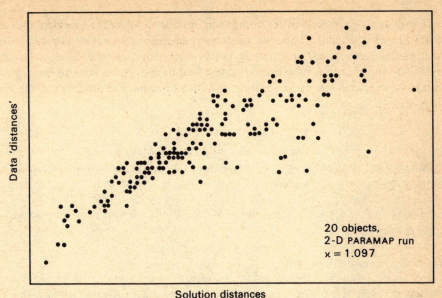

Data 'distances'

Solution distances

20 objects,
2-D PARAMAP run
x = 1.097

Figure 5.3 *Shepard diagram from continuity scaling*

between subjective and geographical distance), it is usually much simpler to interpret the results and predict values outside the current range of data if the transformation is a regular and simple mathematical function. In any event, a monotonic scaling often suggests a simpler, underlying relationship: Shepard functions are often linear or exponential over most of the range of the data. In such cases, having used the more indulgent monotonic assumption, it makes eminent sense to go on to use a more restrictive but simpler transformation and submit the data to metric scaling by a regular transformation.

5.2.3.1 Ratio transformation

The earliest forms of 'metric' multidimensional scaling, dating from the pioneering work of Richardson (1938), assumed simply that data dissimilarities were direct estimates of distances between the points concerned, so that the solution distances are viewed as a ratio transform of the distances of the solution, of the form

$$d_{jk} = b\delta_{jk},$$

where b is the 'proportionality coefficient' or 'scaling ratio', merely allowing for a difference in the actual size of the solution configuration, which is generally considered irrelevant in the MDS context. Given such a set of data, it is a relatively straightforward matter to estimate the dimensionality of the solution space and the co-ordinates of the objects by a method developed by Young and Householder (1941), known subsequently as Eckart-Young factoring (see Appendix A5.2).

5.2.3.2 Linear transformation

Linear transformations preserve information on the equality of intervals or differences, so that if the differences $(a - b)$ and $(c - d)$ are equal in the original data, they will also be equal when transformed linearly.

In many cases, methods of data collection or preliminary scaling yield quantities which clearly are not ratio-level genuine distances, but rather interval-level quantities sometimes referred to as distances. How are such interval-level data to be converted into ratio distances? The use of such distances as data assumes that, at least in the perfect case, the solution distances are a *linear* transformation of the data, that is,

$$d_{jk} = a + b\delta_{jk}.$$

In the usual case, this equation will only hold strictly for the fitted pseudo-distances, that is, $d_{jk}^0 = a + b\delta_{jk}$. We have seen that the proportionality coefficient, b (the scale of the configuration) is arbitrary and merely chosen for convenience. However, estimation of the constant a (the intercept on the Shepard diagram linear regression function)* poses a more serious difficulty referred to as 'the additive constant problem'.

The additive constant problem

The problem can best be illustrated by an example based upon one originally given by Torgerson (1958, p. 403). Consider the matrix of data dissimilarities given in Table 5.2a. It happens that, as they stand, these dissimilarities cannot be represented in Euclidean space. The data do not even all satisfy the triangle inequality axiom of any distance measure (Appendix A2.1). For instance, the axiom requires that $d_{24} \leqslant d_{25} + d_{54}$, whereas in these data $d_{24}(= 6)$ is manifestly *greater* than $d_{25} + d_{54}(= 4)$.

If, however, each dissimilarity in Table 5.2a has a constant value of 2 added to it—that is, if the data are linearly transformed by the equation

$$\delta^{\text{new}} = 2.0 + (1.0)\delta^{\text{old}}$$

then the resulting data matrix is as given in Table 5.2b. It happens that there is a perfect two-dimensional representation of these data given in Figure 5.4. If, however, a constant greater than 2 is added, the data can still be perfectly represented, but only in a space of more than two dimensions.

The linear rescaling problem can be stated as follows: Given a data matrix which may not even be capable of representation in a Euclidean space, can a constant be found (i.e. how can the data be linearly transformed) so that the data can be represented as Euclidean distances (in as few dimensions as possible)?

There is no complete solution to the problem, though several have been proposed, some of considerable complexity (see Messick and Abelson 1956; Cooper 1971). An approach which has proved to be generally adequate is Carroll and Wish's (1973) 'triple equality' procedure (based upon Torgerson (1958, p. 276)) which converts data dissimilarities into distances by application of the 'triple equality difference' (TED) test to estimate the additive constant:

$$a = \max_{i,j,k} (\delta_{ik} - \delta_{ij} - \delta_{jk})$$

The 'triple equality difference' procedure is based upon a very simple idea. Let us

---

*In fact, MRSCAL estimates a slightly different transformation: $d_{jk} = b(\delta_{jk} + a)$, which results in a Shepard diagram where the function goes through the origin.

(a) *Data dissimilarities (relative or comparative distances)*

Object	1	2	3	4	5
1	—	3	4	3	1
2	3	—	3	6	2
3	4	3	—	3	1
4	3	6	3	—	2
5	1	2	1	2	—

$= \delta_{jk}$

(b) *Transformed data (actual distances)*

	1	2	3	4	5
1	—	5	6	5	3
2	5	—	5	8	4
3	6	5	—	5	3
4	5	8	5	—	4
5	3	4	3	4	—

$\delta'_{jk} = \delta_{jk} + 2$

(c) *Triple equality test on data of (a)*

Triple Points	(Max) (i, k) j	Test $(\delta_{ik} - \delta_{ij} - \delta_{jk})$	Result	
(1 2 3)	1, 3 2	4 − 3 − 3	− 2	
1 2 4	2, 4 1	6 − 3 − 3	0	
1 2 5	1, 2 5	3 − 1 − 2	0	
1 3 4	1, 3 4	4 − 3 − 3	− 2	
1 3 5	1, 3 5	4 − 1 − 1	+ 2	(max)
1 4 5	4, 5 1	2 − 3 − 1	− 2	
2 3 4	2, 4 3	6 − 3 − 3	0	
2 3 5	2, 5 3	2 − 3 − 1	− 2	
2 4 5	2, 4 5	6 − 2 − 2	+ 2	(max)
3 4 5	4, 5 3	2 − 3 − 1	− 2	

Additive constant = max $(\delta_{ik} - \delta_{ij} - \delta_{jk})$ = 2

Table 5.2 *Additive constant example*

suppose that the three points (i, j, k) form a straight line in the solution space, such as the line (1, 5, 3) in Figure 5.4a. Then $(d_{ij} + d_{jk})$ will necessarily be equal to d_{ik}, and hence the TED value, which may equivalently be written as $d_{ik} - (d_{ij} + d_{jk})$, will be zero. If j lies *off* the line then $(d_{ij} + d_{jk})$ will be larger than d_{ik} and hence the value of TED will be *negative*. In short, the TED test applied to a set of actual distances will produce a value of 0 for points lying on a line, and a negative value in other cases. Note that in this case the test could here never have a positive value and its maximum value would be zero. The situation is the same when dealing with data or 'relative distances' (where $\delta_{mn} = d_{mn} + a$) except that the TED test will give rise to the value $(0 + a)$ in the case of collinear points and to a smaller value (negative + a) in other cases. Hence the *maximum* value of TED will give the quantity which has to be added to each dissimilarity value to convert it to a genuine distance, i.e. the 'additive constant'. This number may incidentally be

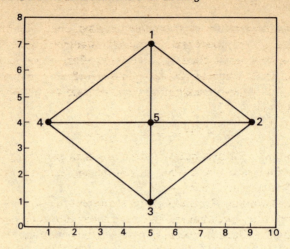

Figure 5.4   *2-dimensional representations of data in Table 5.2b*

negative. As an example, consider the data in Table 5.2. In Table 5.2c an additive constant of 2 is necessary to turn the data into real distances. This value is correctly given by the triples of points (1, 3, 5) and (2, 4, 5), and in both cases the three points lie as a straight line, as can be seen in Figure 5.4a. Even for fallible data, and so long as there are enough points to ensure that at least some triples come close to forming a straight line, this simple method provides an adequate and straightforward way of estimating the additive constant, and is the method used in the INDSCAL program.

5.2.3.3   *Power (and log-interval) transformations*

Power transformations have the general form: $x' = kx^\beta$ and preserve information not only on the equality of intervals—as in the interval scale—but also on the equality of *relative* intervals, i.e. on the ratio of data values. For instance, taking four ratio level data, $a = 3, b = 6, c = 10$ and $d = 20$, then the ratios a/b and c/d both equal $\frac{1}{2}$. When the values are transformed by the power function $x' = 3x^2$, the ratios a/b and c/d are still equal, but now equal $\frac{1}{4}$. In the equation the value of k is an arbitrary factor which cancels out in the formation of ratios; it is the exponent, β, which carries the significant information. Power functions are probably most familiar in the form of compound interest rates in economics and in the 'psychophysical law' in psychology.

A power relationship can always be re-expressed in logarithmic form.* In logarithmic form, power transformations preserve the log *differences* (or intervals) corresponding to the original ratios so that, in the above example, $\log a - \log b = \log c - \log d$ whether with the original values or under the transform $x' = 3x^2$, as can easily be checked. For this reason, the power transformation is sometimes called the logarithmic interval scale, the term adopted in this context by Stevens (1959, pp. 29–30) and Roskam (1972, pp. 495–506). The power transformation is implemented in logarithmic-interval form in the MRSCAL program.

*$10^2 = 100$, and $\log_{10} 100 = 2$, and, in general, if $a^b = c$ then $\log_a c = b$.

The power transformation is a smooth, regular, but non-linear function, illustrated in Figure 5.1(i), whose main parameter of interest is the exponent value which determines how rapidly the slope accelerates. If the power function is drawn in log co-ordinates it then appears as a straight line, with slope equal to the value of the exponent, β. Put in log-interval form, the power transformation† in the case of perfect data would be

$$d_{jk} = a + b(\log (\delta_{jk})).$$

where a represents an additive constant (which may have psychological meaning as the threshold value—see above—but is not usually given substantive interpretation) and b represents the exponent value.

The power transformation has received considerable attention in scaling because of its centrality in early psychological studies of the relationship between physical variables and their subjective counterparts, and also because some data and judgmental processes are known to be best represented by such a transformation.

The 'power law' and its scaling consequences

The work of Fechner and Weber from the 1850s on suggested that human subjects noticed a change in the intensity of a physical variable (such as sound pressure) when the change represented a fixed *proportion* of the previous intensity, i.e. that a *relative* increase in a physical property was perceived as a unit *fixed* increase in psychological intensity. Put slightly differently, the subjective intensity increases as a power function of the physical intensity. Later research has shown that for a wide variety of physical properties, the relationship is well approximated by the so-called psychophysical law (Stevens 1974, p. 361)

$$\psi = k\phi^\beta,$$

where ψ is the perceived magnitude or intensity, ϕ is the physical magnitude, β is the power exponent and k is an arbitrary scaling factor. (In some cases the psychological magnitude only begins to be experienced at a particular threshold and in this case the form of the 'law' needs slight alteration by including an additive constant to represent the threshold effect. Substantive interest focuses the typical value of the power exponent (β) for various modalities.‡

Later experimentation has suggested that a very similar power relationship also exists for the intensity of opinions and attitudes and for the relationship between *direct* estimation (rating) of attitudinal areas and their *indirect* measurement, derived by such methods as Thurstone's law of comparative judgment (Stevens 1966 provides a wide range of examples).

If the 'power law' holds for 'softer', non-experimental and more complex phenomena, as Stevens and others argue it does, then some important consequences follow for scaling studies.

First, 'objective' external properties may well be non-linearly related to scaling solutions based upon subjective or perceptual data. At the very least, it would be

---

†As in the linear case, the MRSCAL program actually estimates: $d_{jk} = b(\log (\delta_{jk}) + a)$.

‡Each modality tends to have characteristic exponent values, ranging from $\frac{1}{3}$ for brightness, $\frac{2}{3}$ for loudness to $3\frac{1}{2}$ for the subjective intensity of electrical current. See Stevens (1974, pp. 362 et seq.).

prudent to allow for this eventuality when engaged upon property-fitting using PROFIT, allowing a 'continuity'-based relationship which will tend to keep increments, and hence ratios, fairly constant, or allowing a monotonic—and hence a power—relationship between the property values and the configuration distances using PREFMAP. In either case it would be foolish only to choose the linear option, which would badly distort a genuine power relationship.

Secondly, the assumption of linearity between the data and the solution is likely to be highly suspect if the data collection method was 'direct' rather than 'derived' (see 2.2 and 2.3). Thus, if the linear transformation is used, it ought to be supplemented by a monotonic fit and/or a 'power' fit, and the Shepard diagrams should be compared.

Thirdly, another way of expressing the power law is that error or variability increases with the magnitude of the data. It is an important consideration in studies of consensus in human judgments (Stevens 1966) and in the development of more recent MDS models, e.g. Ramsay's 'multiscale models', which make explicit assumptions about the likely characteristics of error in the subject's data (see Ramsay 1977, pp. 243–6, especially the discussion beginning with the second paragraph of p. 245, and our section 8.2.1). Perhaps more to the point, if error increases with magnitude it is sensible to pay little attention to dissimilar points in obtaining an MDS solution. This provides a further reason for choosing the local monotonicity or continuity options.

Finally, for some types of data—and especially for confusion data, where the similarity between two objects is taken to be a function of the frequency with which they are confused—there are good theoretical and empirical reasons for expecting an exponential decay (negative power) relationship between the data and the solution distances. Indeed, this same characteristic J-shaped curve has been noted for a goodly number of non-metric scaling studies of co-occurrence frequency data, including Figure 3.14b, and it has been shown that the adoption of a power transformation for the MDS analysis in these circumstances often restores significant local structure which is lost in an ordinal scaling (Arabie and Soli 1977).

5.3   Models

The basic MDS model represents data values as distances. These distances may be thought of as being produced by the combination of latent parameters, i.e. the co-ordinates of the space, which might reasonably be interpreted as scale values along each dimension. It is the particular form of the composition of these co-ordinates to form distances which makes their interpretation as scale values problematical, for we are asserting in the distance model that the scale values for the stimuli are compounded into distances by taking the difference on each scale, squaring it, then summing over each dimension and finally deflating its value by taking the square root. It is possible, however, to regard the data as being linked to a set of scale values or co-ordinates by composition rules other than those of the Euclidean distance formula—and indeed, by variants of the distance formula.

Three major types of composition rule are usefully distinguished:

(i)   *Simple composition.*   Each category of each way of a two-way (or higher) table of data has a scale value, and the composition rule specifies that the entry is

he simple sum of the component categories (the additive model). Other commonly
occurring examples are the difference (a subtractive) and the multiplicative
product) compositions.

(ii) *Scalar product (or factor) composition.*   The objects are located as points
nd or as vectors in a space, and it is the angular separation (scalar product) of the
ectors which corresponds to the data dissimilarities.

(iii) *Distance composition.*   The objects are located as points in a space and the
listance between the points represents the data dissimilarities.

et us take each type of model in turn.

5.3.1 Simple composition

Quite frequently in empirical research the value of a dependent variable is
onsidered to have been produced by the conjoint effect of two or more
ndependent variables or factors, and the researcher is interested in estimating what
he numerical effects are (the scale values) and how they combine. Examples
bound: factorially-designed experiments in agriculture investigate how, and to
what extent, different combinations of soil and fertiliser affect crop yield; social
sychologists interested in impression-formation construct combinations of traits
nd ask subjects to rate the attractiveness of the resulting combinations;
conomists construct portfolios of investments or commodity bundles and ask
espondents to give their preference orderings; demographers calculate the mean
ertility of couples from different regions and occupational groups. In each case the
asic notion is the same: the data are assumed to represent the simultaneous,
onjoint effect of the defining factors, and the purpose of the analysis is to assign a
cale value (estimate a numerical weight) to each constituent category of each
ndependent variable, which, when combined according to the composition rule of
he model, will best fit the values of the dependent variable.

Most researchers will have encountered this type of analysis in the context of
wo (and higher) way analysis of variance and the log-linear analysis of
ontingency tables. In both cases, the underlying model is an additive one: the
alues in the table of the dependent variable are assumed to be the *sum* of the effects
f the relevant categories which define the entry. Given the following 2-way table of
lata, the scale values $A = (8, 2, 5, 4)$ and $B = (3, 1, 6)$ combine additively to
roduce the entries in the table, i.e. $x_{ij} = a_i + b_j$. In most actual applications, an
dditive model will not fit the data perfectly and further interaction terms may need

		B	
	b_1	b_2	b_3
a_1	11	9	14
a_2	5	3	8
a_3	8	6	11
a_4	7	5	10

to be included to represent the unique, joint effect of the categories. However, might be that a transformation—a rescaling—of the data *will* fit an additive model.

In a classic paper, Box and Cox (1964) discuss a number of polynomial transformations of the data, designed to render effects as additive as possible, and in his famous paper Kruskal (1965) developed a procedure based upon monotonic regression designed to find an *ordinal* rescaling of such data which makes them maximally conform to an additive model. The affinity with the basic non-metric MDS model will be obvious, and Kruskal's procedure forms the basis of the additive sub-model of the UNICON program discussed later in the book.

So far we have implicitly assumed that the data form a 2-way table, for purposes of simplicity. Most MDS implementations of simple composition scaling allow up to five such ways, or factors (which may or may not be 'modes', i.e. not distinct sets of objects), although there are few empirical instances of anything more than 3-way table analysis.

Three basic operations form the basis of simple composition models:

(i)   *additive model*: $x_{ij} = a_i + b_j$
(ii) *difference (subtractive) model*: $x_{ij} = a_i - b_j$
(iii) *multiplicative (product) model*: $x_{ij} = a_i \times b_j$

The *additive model* is undoubtedly the best-studied and most used. It turns out to be possible to formulate the necessary and sufficient qualitative conditions that a table must satisfy if it is to be capable of an additive representation. This constitutes a major triumph of axiomatic representationist measurement theory (Krantz et al. 1971, p. 423 et seq.).

The *subtractive model* (or difference model) is appropriate where, for instance, subjects have been instructed to judge the difference between pairs of objects (for example 'imagine 2 different people, each described by one of the adjectives of each pair, and then judge the difference in likeableness between the 2 persons', or where effects are expected systematically to counteract each other).

The *product model* is appropriate where it is thought that categories have a multiplier effect upon each other (or, equivalently, when the logarithm of the effects are additive).

The UNICON program allows the user to define a number of more complex models involving the three simple operations, such as:

$$x_{ijk} = a_i \times b_j + c_k + d_l$$
and
$$x_{ijk} = a_i + b_j - c_k.$$

(See program documentation for details.)

5.3.2 Scalar product models

In the MDS(X) series, all the scalar products (or vector or factor) models assume that the data consist of (or can be reduced to)* a rectangular *two-mode* matrix consisting of a set of (preference) ratings or rankings of a set of p stimuli made by a

*In the MDPREF vector model, input may be a set of pair comparison dominance matrices.

set of N subjects. (In MDPREF this matrix is termed the 'first score matrix'.) For convenience the entries in this matrix are usually denoted s_{ij} to mean the similarity between subject i and object j, or more usually the preference score given by subject i to object j.

The vector solution consists of a configuration of p stimulus points in a user-chosen number of dimensions, and each of the N subjects' set of preference ranks or ratings is represented as a vector, located so that the projections of the stimuli on the vector are in maximum agreement (correlate as highly as possible) with that subject's preferences. The external form of this analysis, i.e. where the stimulus configuration is obtained separately and remains fixed whilst the subject vectors are estimated, was discussed in section 4.4.1.

The purpose of these models is to represent both the stimuli and the subjects in a common 'joint space'. Each subject's preferences are represented as a vector—a projection down, or collapsing of, the stimulus space onto a single dimension—just like the properties embedded in a stimulus space. Interest will chiefly focus therefore on two things:

(i) how well the subject's preferences can be accommodated by the model, and hence represented in the stimulus space (this can be assessed by the correlation of the projections with the original data) and

(ii) how the vectors relate to each other, since the main purpose may be to investigate individual differences in a set of rankings/ratings.

Differences between rankings are signalled in the vector model principally by angular separation. On the one hand, as we saw earlier, the direction in which a vector points is highly significant, for it indicates the manner in which the subject mixes or trades off the characteristics of the stimuli in producing her preferences, and this is measured by the cosine of the angle which the vector makes with the dimensions of the space. By the same token, if we are interested in how one subject vector relates to another, we inspect the angular separation between them—the linear correlation, or cosine of the angle between the two vectors. In inspecting a vector model solution, the first point of interest is how the subject vectors are dispersed around the unit-circle (or sphere).*

If the vector ends are located in a small sector, this indicates high consensus or agreement in subjects' preferences, whereas the more unevenly they are distributed round the circle, the greater the dissensus. The researcher will presumably become interested in whether distinguishably different 'points of view' exist, suggested by small sectors with a high density of vector ends and empty sectors between sectors. If there are different categories of subjects we may also want to know whether the average direction differs significantly between the categories, and statistical tests and procedures for analysing directional data have been developed and are available. (They are discussed in Mardia 1972, and in the MDS context in Coxon and Jones 1979, pp. 128–36 as well as in the MDS(X) documentation for the MDPREF program.)

---

*By convention, subject vectors are normalised to have the same (unit) length in MDPREF. Though this is not a necessary restriction of the model, it makes for greater simplicity if vectors are of standard length. In two dimensions, vector ends will therefore lie around a unit circle, in three (and higher) dimensions, they lie around a (hyper) sphere.

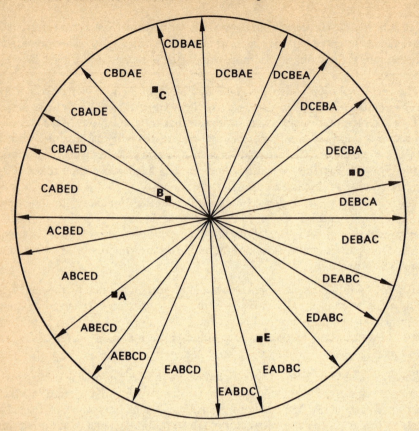

Figure 5.5   *Twenty rankings compatible with 2-D stimulus configurations of 5 points (vector models)*

Although a total of $p!$ (i.e. $p(p-1)(p-2)\ldots \times 1$) rankings of p objects is possible, only a limited number of these can be accommodated within a stimulus configuration. We therefore need to enquire both how many rankings can be accommodated in a configuration of p points in r dimensions and how they are represented therein. As an example, take the 5-point stimulus configuration given in Figure 5.5. There are 5! = 120 possible rank orderings of 5 stimuli, but only 20 of these can be represented perfectly in a given vector configuration of 5 points in two dimensions, and one half of these will simply be mirror-images of each other formed by reversing the direction of the vector. The 20 rankings compatible with this configuration are given in the figure. Notice that there is an orderly interlocking between the rankings, akin to that shown by Coombs (1964, p. 87 et seq.) in the context of discussing the unidimensional unfolding (distance) model for preferences. As one moves around the circle, only adjacent stimuli are interchanged in the rankings (beginning in the north-easterly position and moving clockwise: *DCBEA, DCEBA, DECBA, DEBCA*, and so forth).

Although the scalar products model has been described as a point (stimulus) and vector (subject) representation, formally the model is expressed entirely in terms of vectors—a set of vectors drawn from the origin of the space to the location of each

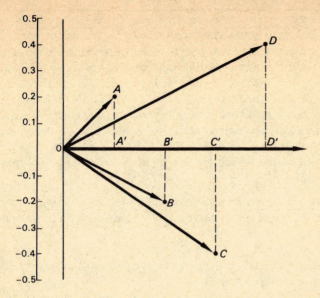

Subject vector y	Stimulus point x	Scalar product = yx'	Projection
(1,0)	A (0.2, 0.2)	0.2	OA' = 0.2
	B (0.4,−0.2)	0.4	OB' = 0.4
	C (0.6,−0.4)	0.6	OC' = 0.6
	D (0.8, 0.4)	0.8	OD' = 0.8

Figure 5.6   *Projections and scalar products*

stimulus, and a set of unit-length *subject* vectors. The key to understanding the formula for the model is knowing that the scalar product of the stimulus vector with the (unit-length) subject vector is the same as the vertical projection of that stimulus point on the subject vector. This property is illustrated in Figure 5.6, where the subject vector is drawn along the first dimension to simplify the arithmetic.

Let $\mathbf{x}_j$ represent the vector from the origin to the location of stimulus j in r-dimensional space, and $\mathbf{y}_i$ represent the (unit-length) vector for subject i, then the preference value which stimulus j has for subject i is estimated as the scalar product of the vector concerned:

$$\hat{s}_{ij} = \mathbf{y}_i \cdot \mathbf{x}'_j = \sum_{a=1}^{r} y_{ia}x_{ja}$$

or in matrix form:

$$\hat{\mathbf{S}} = \mathbf{YX}'$$

The matrix of preference scores estimated by the model is termed the 'second-score matrix', and the purpose of the vector model is to obtain a stimulus configuration X and subject vectors Y, so that the discrepancy between the original 'first-score' data (s_{ij}) and the estimated 'second-score' values $(\hat{s}_{ij})$ is as small as possible. (In the case of a non-metric version, the monotonically transformed data will be compared to the estimated values.) Carroll (1972, p. 124 et seq.) and the MDS(X) documentation describe the stress-like index of agreement, C_1 used to measure the goodness of fit. The method of solution involves factoring two product matrices formed from the first score matrix.*

The main properties of the vector model (cf. Roskam 1968, p. 28) may be summarised as follows:

(i) *Increasing utility*. A subject's preference (or similarity rating) increases continuously in the direction of the vector; the further out an object projects on it, the more it is preferred.

(ii) *Mediocrity*. An object may always occupy a position between the extremes of all the subject's preferences, i.e. never be either most or least preferred (see object B in Figure 5.5, for example).

(iii) *Reversability*. If a given ordering occurs, the opposite ordering may also occur. Indeed, the orderings compatible with a given stimulus configuration divide into two opposite halves, producing the characteristic 'spokes of a wheel' isotonic regions (sector of the space where the same rank ordering of stimuli is implied) seen in Figure 5.5.

The vector model differs considerably in these respects from the distance (unfolding) model of preference discussed in the next section. The differences and the related issues of interpretation of configurations produced by programs implementing the models are discussed in Chapter 6.

5.3.3 Distance models

The central idea of distance models is that the proximity of points in a space is used to represent their empirical similarity, or equivalently that distance represents their dissimilarity. In the vast majority of MDS models, the distance function involved is the familiar Euclidean form, but Euclidean distance is only one special case of a whole family of distance functions, each with its own characteristics and properties (see Appendix A2.1.1.2). Proceeding from the familiar to the less familiar, we shall discuss the basic distance model first, then move on to look in greater detail at the properties of Euclidean and other types of distance.

Given a set of distances it is always possible to reconstruct the configuration of points which generated them. (This procedure is described in Appendix A5.2.2 and forms the basis of classic metric scaling discussed above.) However, such a recovered configuration is not unique, in that several aspects of it are arbitrary and

---

*The first score matrix S is approximated in the user-chosen dimensionality a, by a least squares approximation $S = YX'$ (of rank a) using the Eckart-Young factorising procedure. The eigenvectors of the minor product matrix $S'S$ provide estimates of the stimulus configuration Y, and the eigenvectors of the major product matrix SS' provide the estimates of the subject vectors X, when the rows are normalised to unity. The eigenvalues of both product matrices are the same and indicate the concentration of variation in the principal axes (see Appendix A5.2.2).

may be changed at will. (These have been mentioned before (4.1), and are further discussed in Appendix A7.1.) In particular, the actual size or scale of the configuration and the origin of the space are arbitrary. Moreover, the orientation of the axes may be changed and reflected at will. Strictly speaking, it is only the relative distance between points which is significant in interpreting a distance model solution—the origin and axes simply provide a convenient framework to locate the points.

5.3.3.1   Point-point (two-mode 'unfolding') distance models

When the data consist of a rectangular two-mode matrix, of rankings or ratings, then the distance model can be used to represent both the stimuli *and* the subjects as points. The solution consists of a configuration of p stimulus points and N subject points where each subject is represented as being at a 'maximal' or 'ideal' point, located in such a way that the distances from this point to the stimulus points are in maximum agreement with the subject's preference ratings or rankings.

In external models such as PREFMAP phase III, the stimulus configuration is obtained separately and remains fixed whilst the 'subject' or property points are estimated (see 4.4.2), whereas in internal models, such as MINI-RSA, both sets are estimated simultaneously. As in the case of the vector model, both metric and non-metric versions exist—in the former a linear correlation between the preference data and the subject-stimulus distances is maximised while in the latter a variant of stress involving only the rank order of the data is minimised.

The position of the 'ideal point' is interpreted as the one point in the space where the subject's preferences are at a maximum, and her preference decreases in every direction. This is often termed a 'single peaked preference function', since it assumes that there is only *one* point of maximum preference.

The non-metric version of the distance model is best known under the title of 'unfolding analysis', developed by Coombs (1964, chs. 5–7). The two-dimensional case is illustrated in Figure 5.7 with reference to the same 5-stimulus configuration used in the vector model case (Figure 5.5).

A midline is drawn between each pair of points, dividing the space up into 46 isotonic regions. Every ideal point within one of these regions possesses the same rank order of distances to the five stimuli. This is illustrated in Figure 5.7; thus in region I the corresponding I-scale is ABECD, and in crossing over the midline CE to region II, the I-scale becomes ABCED. Similarly the move from region III to IV represents the transition from DBCEA to DBECA. Notice that some regions are entirely encompassed by midlines (closed isotonic regions), whilst others at the periphery are not (open isotonic regions). Herein is an important distinction between the vector and distance models: the vector model excludes closed regions (see the corresponding Figure 5.5) and can accommodate fewer I-scales than the distance model. The maximum number of I-scales compatible with the two models is illustrated below in Table 5.3 (see Coombs 1964, Tables 7.1 and 12.9).

Normally the points corresponding to the most popular or consensual rankings will lie at the centre of the space, and the least popular ones at the periphery. Research has shown, as Coombs originally suggested, that ideal points within the 'open' isotonic regions are located with less accuracy than those in the closed ones. Moreover, the fewer the midlines constraining a region, the more likely it is that the

No. of dimensions

No. of points	2		3		4		5		Total possible (p!)
	D	V	D	V	D	V	D	V	
1	1	1	1	1	1	1	1	1	1
2	2	2	2	2	2	2	2	2	2
3	6	6	6	6	6	6	6	6	6
4	18	12	24	24	24	24	24	24	24
5	46	20	96	72	120	120	120	120	120
6	101	30	326	172	600	480	720	720	720
7	197	42	932	352	$2.56_{10}3$	1512	$4.32_{10}3$	$3.60_{10}3$	$5.04_{10}3$
8	351	56	$2.31_{10}3$	646	$9.08_{10}3$	$3.98_{10}3$	$2.22_{10}4$	$1.42_{10}4$	$4.03_{10}4$
9	583	72	$5.12_{10}3$	$1.09_{10}3$	$2.76_{10}4$	$9.14_{10}3$	$9.49_{10}4$	$4.60_{10}4$	$3.63_{10}5$
10	916	90	$1.04_{10}4$	$1.74_{10}3$	$7.36_{10}4$	$1.90_{10}4$	$3.43_{10}5$	$1.28_{10}5$	$3.63_{10}6$

MODEL: DISTANCE/VECTOR

Table 5.3 Total possible number of I-scales, and totals compatible with the distance and vector models for p points in r dimensions)

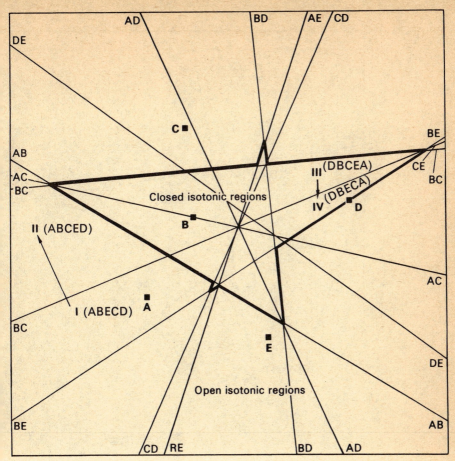

Figure 5.7   *Forty-six rankings compatible with 2-D stimulus configuration of 5 points (vector model)*

subject point be mislocated in a scaling solution. This is well illustrated in Figure 5.8, representing the scaling of the 18 I-scales compatible with a 2-dimensional, 4-point configuration (Coombs, 1964, Figure 7.4, p. 146). Each small square represents the position of a subject as located by the relevant non-metric program (MINIRSA). Note that in the case of the closed regions, the squares are all located within the correct region, although they are deflected to the outer edge. In the case of the open regions, those defined by three lines are correctly located near the centre of the region but those defined by only two lines are, without exception, displaced slightly outside their correct location.

The multidimensional unfolding model is hence clearly more 'tolerant' than the vector model, in the sense that it can accommodate more I-scales (see Table 5.3). So long as the number of stimulus points is large compared to the number of dimensions, the size of the isotonic regions is small, especially towards the centre of the configuration, and they become increasingly well-represented by a point. For this reason, stimuli points in the central part of a configuration are normally the most stable, whilst those at the periphery can usually be moved around fairly freely without affecting the goodness of fit. The variation in judgments about particular

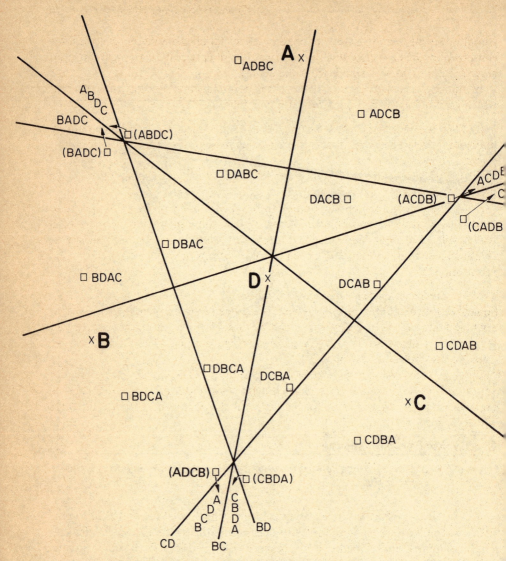

Stress$_2$ = 0.0003 after 45 iterations

N.B. Arrows indicate discrepancy between 'true' locations
and scaled ('recovered') location

Figure 5.8   *Actual and scaled location of isotonic regions*

stimuli is also an important factor in assessing the stability of a configuration in an
internal scaling model. Highly popular stimuli will tend to be projected into the
centre of the subject points (so that they can feature close to most subject's ideal
points) and highly unpopular stimuli will be located at the outside of a
configuration. Indeed, if a stimulus is sufficiently unpopular it can be located
virtually anywhere on the periphery, so long as it is at a maximum distance from
the ideal points. An example of this occurs in the analysis of the Delbeke data
reported in section 6.2.2 (see Coxon 1974) where virtually all subjects rejected the

stimulus 'no children' in a study of preferences for families of different sizes and sex composition. When scaled, this stimulus was located at greatly varying points, but always at an extreme distance from the centre.

In summary, the properties of the point-point (distance) model of preference which contrast with the vector model are as follows:

(i) *Single peakedness.* It is assumed that each subject has one single point of maximum preference and that preference decreases (symmetrically) from this point.

(ii) *Excellence.* If the distance model holds, then each stimulus must be preferred most by at least one subject.

(iii) There is nothing corresponding to the reversability property of the vector model in the multidimensional unfolding model: some mirror-image pairs of I-scales will exist, but not others. More importantly, the distance model is characterised by the presence of closed isotonic regions, which cannot occur in the vector model.

5.3.3.2 Euclidean and non-Euclidean distance

So far, 'distance' and 'Euclidean distance' have been used interchangeably. In fact, a whole family of distance measures can be defined for a given configuration of points. Our interest shifts away from the correct location of points to how we measure the distance between them.

Three types of distance have been found useful in MDS and are represented in various MDS(X) programs: city block, Euclidean and dominance metrics. These are all special instances of the Minkowski *r*-metric family of distance measures which have the form:

General (Minkowski) Distance

$$d_{jk}^{(r)} = r\sqrt{\sum_a |x_{ja} - y_{ka}|^r}$$

where x_{ja} is the co-ordinate of the *k* th point and y_{ka} is the co-ordinate of the *j* th point on the *a* th dimension and *r* is the Minkowski *r*-metric power.

Each value of *r* (between 1 and infinity) defines a distinct metric distance. Each can be thought of as a simple composition model—a 'powered additive difference' model which asserts (Beals et al. 1968 pp. 133–5) that:

(i) absolute *differences* on each dimension, *a*
(ii) which are raised to the same *power r*
(iii) combine *additively* over the dimensions to produce
(iv) the overall distance between a pair of points, *j* and *k*.

In the case of Euclidean distance, the power is 2, so differences are squared, and the final distance measure deflates the value by taking the square root.*

*Carroll and Wish 1974, p. 412 et seq. argue persuasively that the final *r*-th root may often be usefully ignored, and when this is done a wider range of models qualify as metrics. In the Euclidean case, a number of models are more simply expressed and best understood by treating *squared* distances (i.e. ignoring the final square root). Carroll and Wish (ibid. p. 413) and Shepard (1974, p. 405 et seq.) discuss even more general distance measures, some of which do not even satisfy the triangle inequality.

Euclidean Distance

$$d_{jk} = \sqrt{\sum_a (x_{ja} - x_{ka})^2}$$

where x_{ja} is the co-ordinate of the j th point and x_{ka} is the co-ordinate of the j th point on the a th dimension.

The three commonly-used types of distance mentioned above are illustrated in Figure 5.9. The basic difference lies in the question of whether the differences between objects on each dimension remain separate or merge together ('interact') in producing the overall distance. For $r = 1$ (city block metric) all the dimensional

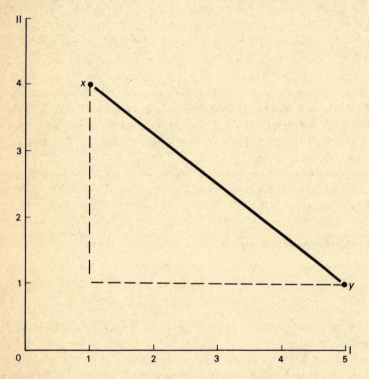

General Minkowski r–metric

$$d_{xy}^{(r)} = \sqrt[r]{\sum_a |x_a - y_a|^r}$$

City block metric ($r = 1$)   (dashed lines)

$$d_{xy}^{(1)} = \sum_a |x_a - y_a| = 4 + 3 = 7$$

Euclidean metric ($r' = 2$)   (solid line)

$$d_{xy}^{(2)} = \sqrt[2]{\sum_a |x_a - y_a|^2} = \sqrt{4^2 + 3^2} = 5$$

Dominance metric (approximated by $r = 32$)

$$d_{xy}^{(32)} = \sqrt[32]{\sum_a |x_a - y_a|^{32}} = \sqrt[32]{4^{32} + 3^{32}}$$

$$= \sqrt[32]{1.8447_{10}19 + 1.8530_{10}15}$$

$$= \sqrt[32]{1.8449_{10}19}$$

$$= 4.0000125$$

Figure 5.9   *Minkowski metrics*

differences have the same weight in determining the distance; they are simply added together. As r goes to infinity (dominance metric) the largest single difference comes to swamp out all other information. By contrast, the Euclidean distance can be thought of as a compromise where no dimension has a specially important status.

The Euclidean metric is the only one where the orientation of the axes is arbitrary, in the sense that a rotation will leave the distances unchanged. *In all other Minkowski metrics the distances are defined by reference to a fixed set of axes and any rotation will change the distance values.* It is for this reason that axes should be drawn in any configuration where the distance is non-Euclidean.

This property is illustrated by the Minkowski unit-distance (iso-similarity) contour diagram in Figure 5.10. More complex variants are given in Roskam (1968, p. 51) and in Carroll and Wish (1974, p. 417). If all the points at a fixed distance from the origin of a 2-dimensional space are joined, then they form a circle in the case of Euclidean distance (the circle defines the equation $p^2 + q^2 = r^2$, which in this case corresponds to $(x_1 - y_1)^2 + (x_2 - y_2)^2 = d_{xy}^2$ of Figure 5.9). Wherever the dimensions are rotated, the squared dimensional differences still total one, so all are equally permissible. In the case of city block distance, the points at a fixed distance from the origin form a diamond (the diamond is defined by the equation $p + q = r$, corresponding to $(x_1 - y_1) + (x_2 - y_2) = d_{xy}$ of

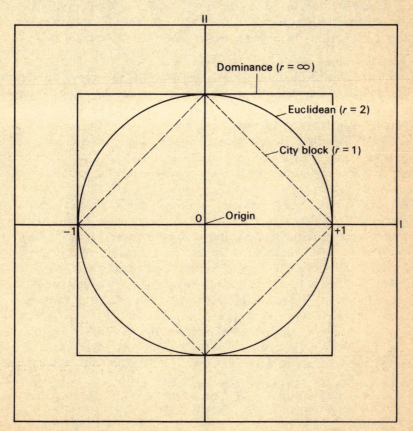

Figure 5.10 *Equal distance contours for 3 Minkowski metrics*

Figure 5.9). If the axes are rotated through anything other than 90° (or multiples of it) the sum of the differences will no longer be the same. For the dominance metric, the unit contour is a cube: until the two differences become equal, only the larger makes any contribution to the distance. Once again, a rotation of anything but multiples of 90° will destroy this relationship.

There is persuasive psychological and empirical evidence (Attneave 1950, Torgerson 1958, Hyman and Well 1968) that the city block metric is particularly appropriate where the characteristics of the objects are obviously compelling or perceptually distinct. By contrast, where the characteristics are more complex the dimensional information begins to merge or blur, and the Euclidean distance will provide a better description. (Compare judging pairs of triangles differing in size and orientation to judging towns in terms of their desirability.) Arnold (1971) has argued that the dominance model provides a better account of data collected by procedures which impose heavy information processing demands on the subject, although the analysis has been questioned by Carroll and Wish (1974).*

A good deal of evidence underlines the conclusion unequivocally argued by Shepard (1974, p. 407) and Carroll and Wish (1974, p. 420) that the Euclidean metric appears to be robust against even extreme departures from its assumptions. Moreover, city block and dominance metrics turn out to be rather subject to local minima and degenerate solutions (information on 8 points can be fit perfectly, in a totally degenerate way, in 3 dimensions, see Shepard 1974). Even if users wish to scale in a 'simpler' metric they are advised to begin with a Euclidean solution and work down (or up) to the preferred metric (Arabie 1973, Shepard 1974).

5.3.3.3   Generalised distance and other metrics
Three other types of distance occur in the MDS(X) programs. The first type (weighted Euclidean distance, a generalisation of Minkowski metrics) is employed in analysing three-way data and is dealt with in the next chapter.

The other two types of distance are simpler than the Minkowski metric and neither are necessarily capable of being represented in a dimensional space. The humblest is simply a metric that obeys the triangle inequality, and the other is the hierarchical clustering or tree-metric that obeys the somewhat more stringent ultra metric inequality.

The simplest type, 'non-dimensional scaling', relaxes not only the additivity requirement of Minkowski spaces, but also the minimum dimensionality of dimensional smallest-space analysis. It does so by dispensing entirely with the idea of a co-ordinate space as embedding the distances, and seeks instead to rescale the data dissimilarities into a set of distances which perfectly obey the triangle inequality, and are as close as possible to being a monotone function of the data. This is achieved by a process of successively increasing the variance of the distances. Intuitively this can be likened to the conformal mapping discussed in 5.2.2, where the largest distances are increased and the smaller ones decreased. This has the effect of forcing down the dimensionality of a space—as in conformal mapping down from a sphere onto a flat plane. In this non-dimensional case, the

*Koopman and Cooper (1974) rightly stress that in two dimensions it is impossible to tell *mathematically* whether the city block or dominance metric is appropriate, since the one is a mathematical transformation of the other—indicated by the fact that the unit contours simply represent a 45° rotation of each other.

analogy does not hold exactly, but it produces 'better behaved' distances. This process is known as maximum variance non-dimensional scaling (Cunningham and Shepard 1974; implemented in MDS(X) as MVNDS) and described in 6.1.7.

The chief virtue of the model is its generality and simplicity: all the other models are special, more restrictive versions of it and only minimal assumptions have to be made to obtain a solution. By dispensing with the assumption of an underlying continuous space, it may also be possible to find a better, more law-like relationship between the original and the rescaled data. Moreover, for the cautious user, this procedure could be used as the first part of the scaling process: obtain a good estimate of the shape of the monotone function without assuming any particular Minkowski metric, and the resulting distances can then be used as input for a more restrictive distance model of one's choice—thus avoiding the dangers of degeneracy and local minima to which non-Euclidean distance scaling is prone. Alternatively, the user might decide to represent the rescaled data in some other way: as a graph, or a tree (i.e. as input to a clustering program).

The other type, the tree-metric, defined by the ultra-metric inequality, was encountered earlier in section 4.3.3.1 as the defining characteristic of a hierarchical clustering scheme (HICLUS program). If a set of data obeys this criterion it can be represented as a dendogram (or rooted tree) where the distance between any two points is defined as the level at which they join (see Figure 4.2).

APPENDIX A5.1 KAPPA AND RELATED MEASURES OF DISCONTINUITY

All Shepard-Carroll (1966) kappa-based measures of continuity have the basic form:

$$\text{kappa} = \left[\left(\begin{array}{c}\text{smoothing}\\\text{ratio}\end{array}\right) \times \left(\begin{array}{c}\text{local proximity}\\\text{weight}\end{array}\right)\Bigg/\left(\begin{array}{c}\text{normalising}\\\text{factor}\end{array}\right)\right]$$

$$\qquad\qquad\qquad\text{(i)}\qquad\qquad\qquad\text{(ii)}\qquad\qquad\qquad\text{(iii)}$$

(i) *Smoothing*

The basic notion of a 'smooth transformation' consists in comparing two uni-dimensional continua, $\mathbf{x}$ and $\mathbf{y}$, in terms of a mapping or transformation which ensures that the intervals or differences between adjacent points (i, j) in the one are of approximately the same size as those in the other, i.e. that the interval $(y_i - y_j)$ is of about the same size (apart from differences in scale) as the interval $(x_i - x_j)$. This is achieved by studying the *ratio* of the differences for each adjacent pair, squaring the result for purposes of convenience. Thus,

$$\left(\frac{\Delta y}{\Delta x}\right)^2 = \left(\frac{y_i - y_j}{x_i - x_j}\right)^2 \tag{1}$$

In the case of MDS, where the data 'distance' ($\mathbf{y}$) are being compared to the solution distances ($\mathbf{x}$), the differences in (1) become distances, and a simple overall measure of discontinuity or 'lack of smoothness' between the data and the solution is formed by summing the ratio over all $p(p - 1)/2$ pairwise data points:

$$\sum_{i \neq j}\sum \left(\frac{\delta_{ij}^2}{d_{ij}^2}\right) \tag{2}$$

(ii) Local proximity weight

Local proximity weights w_{ij} are the extension to the multidimensional case of the restriction to adjacent pairs (representing changes in value) in the uni-dimensional case. Shepard and Carroll (1966, p. 582) show that only weights having the form

$$w_{ij} = d_{ij}^s, \quad \text{with} \quad s < 0$$

can ensure both that local monotonicity is enforced ($s < 0$), and that the ratio of any two weights remains invariant under change of scale, since solutions are unique only up to similarity transforms. As they indicate, simplicity and experience show that $s = -2$ is a sensible choice, yielding weights of the form: $(1/d_{ij}^2)$, which when multiplied into the discontinuity ratio (1), yields the basic measure of discontinuity,

$$\text{raw kappa} = \sum_{i \neq j} \frac{(\delta_{ij}^2)}{d_{ij}^4} \tag{3}$$

(iii) Normalising factor

Shepard and Carroll (1966, p. 582) define a normalising factor which ensures that kappa reaches a minimum when the solution distances d_{ij}^2 are proportional to the data distances δ_{ij}^2 except for a similarity transform. The simplest such factor is

$$\sum_{i \neq j} (1/d_{ij}^2)^{-2}. \tag{4}$$

The product of (3) and (4) yields the basic (normalised) kappa index. Gower (1979, p. 3) shows that this normalised kappa measure can be written in a particularly simple and interpretable form:

$$\text{normalised kappa} = \sum_{i \neq j} w_{ij} \left(\frac{1}{\delta_{ij}^2} - \frac{1}{d_{ij}^2} \right) \tag{5}$$

If the d_{ij} and δ_{ij} are of approximately the same order of magnitude, the weighting factor is approximately equal to $(1/d_{ij}^6)$—which emphasizes fairly starkly how drastic a weighting function it is, giving long distances virtually no influence in determining the final configuration, and giving short (proximate) distances enormous weight. Even small differences in short distances will have very considerable effect on the size of the kappa measure; many users may prefer a less punitive weighting factor.

A5.1.1 Generalised forms of continuity index

Normalised kappa is a special case of the family of continuity indices referred to as 'kappa star'

$$\kappa^* = \sum_{i \neq j} \frac{(\delta_{ij}^2)^a}{(d_{ij}^2)^b} \bigg/ \left[\sum_{i \neq j} (d_{ij}^2)^c \right]^{-b/c} \tag{6}$$

(Normalised kappa is the case where $a = 1$, $b = 2$ and $c = -1$.) If the normalising factor is to keep the kappa index invariant under a similarity transform on the solution space, then the exponents must satisfy the condition $b + c - a = 0$, and c should be negative. (The exponent values can be varied within the PARAMAP program, where the default values produce the normalised kappa index.)

In terms of the components of the index, a and b affect the continuity ratio, b (and, more indirectly, a) affects the strength of the local monotonicity weight, and b and c affect the normalising factor. Kruskal and Carroll (1969) have argued for $a = b = 1$, thus minimising the local monotonicity weighting and making all 'changes' and distances of equal importance in the minimisation process. They also make the case for reducing the size of the exponents, suggesting two further possibilities:

(i) $a = \frac{1}{2}$, $b = 1$, when the ratio takes the especially simple form of (δ_{ij}/d_{ij}^2), which still preserves local monotonicity weighting, but not in a way that so severely reduces the effect of larger distances; and

(ii) $a = b = \frac{1}{2}$, which removes the local monotonicity weighting and concentrates the effect on the simple ratio of the two distances (δ_{ij}/d_{ij}).

In general, it is necessary that $b > a$ if local monotonicity is to be maintained: the greater the inequality, the more severely discrepancies in representing local structure are penalised, and the less the balancing effect of more distant points.

APPENDIX A5.2   CONVERTING DISTANCE INTO SCALAR PRODUCTS AND BASIC CLASSICAL SCALING

A5.2.1   Conversion of distances into scalar products

In Appendix A2.1 it is shown how to convert scalar products into Euclidean distances. The reverse is often more useful and necessary—how to turn distances into scalar products. This forms the initial stage of most classic metric scaling procedures and is also often used to produce an initial configuration in non-metric models.

(i)   *Converting distances into scalar products**

We assume that the distances are genuine and not relative or 'errorful' distances, and to simplify matters we shall assume we are dealing with *squared* distances. Then the required conversion formula is as follows:

$$b_{jk} = -\tfrac{1}{2}(d_{jk}^2 - d_{\cdot j}^2 - d_{k\cdot}^2 + d_{\cdot\cdot}^2) \tag{1}$$

where b_{jk} is the scalar product between vectors j and k,

$$d_{\cdot j}^2 = \sum_k^n d_{jk}^2/n, \; d_{k\cdot} = \sum_j^n d_{jk}^2/n \quad \text{and} \quad d_{\cdot\cdot}^2 = \sum_j\sum_k d_{jk}^2/n$$

and n is the number of distances.

Formula (1) can be derived easily from the definition of Euclidean distance so

---

*This section relies on Carroll (1973). Alternative derivations will be found in Torgerson (1958, p. 255 et seq.).

long as we are dealing with genuine distances (rather than relative or 'errorful' ones) and if we simplify the arithmetic by placing the origin of the space at the centroid of the points, and deal with *squared* distances rather than the distances themselves.

By definition:

$$d_{jk}^2 = \sum_a (x_{ja} - x_{ka})^2 \tag{2}$$

which when multiplied out gives

$$d_{jk}^2 = \sum_a (x_{ja}^2 - 2x_{ja}x_{ka} + x_{ka}^2)$$

$$= \sum_a x_{ja}^2 - 2\sum x_{ja}x_{ka} + \sum x_{ka}^2 \tag{3}$$

This first and third terms on the right are the squared norms of j and k respectively (the vector drawn from the origin to the points concerned), denoted l_j and l_k, hence:

$$d_{jk}^2 = l_j^2 + l_k^2 - 2\sum x_{ja}x_{ka} \tag{3a}$$

The cross product term on the right of (3a) corresponds to the scalar product between vector j and k, denoted b_{jk}, so the last equation can be simplified and rewritten as

$$d_{jk}^2 = l_j^2 + l_k^2 - 2b_{jk} \tag{4}$$

Since l^2 is defined as the averaged squared distance from the origin (i.e. $\sum_j l_j^2/n$) then the following equalities can be shown to hold:

$$d_{.k}^2 = l^2 + l_k^2; \quad d_{j.}^2 = l_j^2 + l^2 \quad \text{and} \quad d_{..}^2 = 2l^2$$

(where the dot signifies the average over the relevant subscript). Substitution in (4) yields

$$d_{jk}^2 = d_{.k}^2 - d_{j.}^2 + d_{..}^2 = -2b_{jk}$$

which can be re-arranged as

$$b_{jk} = -\tfrac{1}{2}(d_{jk}^2 - d_{.k}^2 - d_{j.}^2 + d_{..}^2) \tag{5} = \text{(1)}$$

thus yielding the necessary conversion formula.

As an example, let us return to the distance matrix used in Figure A2.1:

$$\mathbf{D}^2 = \begin{bmatrix} 0 & 5 & 25 \\ 5 \cdot & 0 & 8 \\ 25 & 8 & 0 \end{bmatrix}$$

Let us calculate the scalar product b_{32} by (5):

$$b_{32} = -\tfrac{1}{2}(d_{32}^2 - d_{.2}^2 - d_{3.}^2 + d_{..}^2)$$

$$= -\frac{1}{2}\left(8 - \frac{13}{3} - \frac{33}{3} + \frac{76}{9}\right)$$

$$= -\tfrac{1}{2}(1.11) = -0.56$$

which is precisely the scalar product (calculated from the centroid as deviate scores) b_{32} given in Appendix A2.1.2.

The conversion formula as can be seen in (5) involves 'double-centring' the squared distance matrix, i.e. removing the row effects, the column effects and adding back in the grand mean.

A5.2.2 The scalar products matrix B and classic scaling

The scalar product matrix, **B**, has a number of properties which are crucial to recovering the space which generated the original distance. Young and Householder (1941) showed that:

(i) If **B** is positive semi-definite (Gramian)—as is necessarily the case if we are dealing with real distances—then by definition its latent roots will be non-negative. This means that the distances can be represented in a real Euclidean space.

(ii) The rank of **B** is equal to the number of dimensions necessary to represent the distances.

(iii) **B** can be factored by conventional methods to obtain a matrix **A**:

$$\mathbf{B} = \mathbf{AA}'$$

where **A** is a matrix whose elements (a_{ij}) give the projection or co-ordinates of stimulus i on the j th dimension. (These co-ordinates are only unique up to a similarity transform).

Moreover, Eckart and Young (1936) show that if one wishes to obtain a solution in as small a dimensionality as possible (i.e. to approximate a full solution of r dimensions by one in $q \ll r$ dimensions), then the corresponding matrix of co-ordinates (call it **C**, of order q) which minimises the sum of squares of the difference between the full and the approximate solution is given by

$$\mathbf{C} = \mathbf{A}^* \, \varLambda \, \mathbf{A}^{*\prime}$$

where $\varLambda$ consists of the first q latent roots of **B** (in order of magnitude), and **A**** (an incomplete version of **A**) consists of the corresponding q columns or latent vectors of **A**. (see Torgerson 1958, p. 255 et seq. and van de Geer 1971, p. 70 et seq.).

These theorems of Young, Householder and Eckart provide a straightforward way to recover the space that generated a set of distances and produce a close-fitting approximation in a lower dimensionality. The first is achieved by turning distances into scalar products by applying formula (5) and then factoring the resulting matrix to obtain the co-ordinates, which will be unique up to a similarity transformation, and the second is achieved by restricting attention to the first q latent vectors of the matrix.

But these procedures only hold if the data are genuine distances; if they are only distance estimates' or relative distances, then we shall encounter the additive constant problem discussed in 5.2.3.2.1 above. Nonetheless, this classic scaling solution turns out to be remarkably robust, and forms an integral part of obtaining the initial configuration for non-metric models, of the now more sophisticated two-way distance metric scaling models and in the basic three-way model, INDSCAL.

6 Programs For Analysing Two-Way Data

Now, in perusing what follows, the reader should bear in mind not only the general circuit as adumbrated above, with many sidetrips and tourist traps, secondary circles and skittish deviations but also the fact that far from being an indolent partie de plaisir, *our tour was a hard, twisted teleological growth whose sole* raison d'être *(these French clichés are symptomatic) was to keep my companion in passable humour from kiss to kiss.*

V. V. Sirin, *Taina* (p. 260)
translated from the Russian by
Vivian Darkbloom as *Tatyana* (p. 19)

In this chapter programs implementing models and transformations appropriate for analysing two-way data are described (further details are contained in the *MDS(X) User's Manual* Reports).

The defining characteristics of the relevant MDS(X) programs are described in Table 6.1, and the sections of this chapter follow the sequence indicated in the final column of the table. For two-way data, the basic distinction is whether the solution represents the relationship between *one* set of objects, the stimuli (one-mode) or between *two* sets of objects, the subjects and the stimuli (two-mode). This distinction is used to organise this chapter.

The chapter is written in such a way that the user can either read it through sequentially to get an overview of all the relevant programs, or else go directly to the section (program) which is of most immediate interest. Each section is therefore largely self-contained and refers to a single program or two highly similar programs.

The structure of each section is also the same and consists of:

a brief definition of the program, listing its specific characteristics;
an extended description of the program and its uses; followed usually by
an example of its application.

6.1 One-mode Data

A mode refers to a single class of entities, which could in fact be stimuli, subjects, test items, occasions, geographical sites, botanic species, lexical items etc. (Carroll and Arabie 1980). Two-way single-mode data consist typically of a matrix giving (dis)similarity or other proximity measure between each pair of objects taken from the class of entities. Such data are certainly the most common form of data.

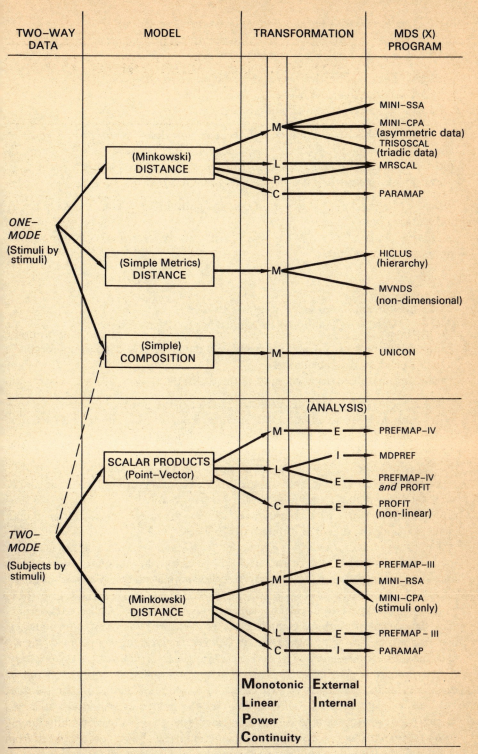

Table 6.1 *Analysis of 2-way data by MDS(X) programs*

6.1.1 *The basic non-metric model* (MINISSA(N) *and* SSA(M))

Concisely: MINISSA (Michigan-Israel-Nijmegen Integrated Smallest Space Analysis) in its *Nijmegen* and *Michigan* versions provides

internal analysis of a two-way symmetric matrix of (dis)similarities
by a Euclidean distance model
using a monotone transformation of the data.

The form of this basic non-metric model has been described fully in 3.5 and 3.6. The simplest and most efficient implementation is the Nijmegen version (MINISSA) and there is also the more extended Michigan version (SSA(M)) which contains parameters for permitting a wider range of alternatives:

Data Missing data, fixing part of the configuration (in effect, external scaling) and adding points
Transformation Local or global monotonicity, Kruskal or Guttman's minimisation procedure.

6.1.2 *Conditional Proximity Analysis* (MINICPA)

Concisely: MINICPA (Conditional Proximity Analysis) provides:

internal analysis of two way (dis)similarity data in a row-conditional format
by a Euclidean distance model
using a monotone transformation of the data.

The MINICPA program is designed to implement the third approach to scaling asymmetric data described in 5.1.1.1. It differs from MINISSA only in that each row of the square data matrix is assumed to be conditional, i.e. as a rank order in its own right. Usually, then, δ_{jk} will not normally equal δ_{kj} and each of the two dissimilarities are fit by a separate disparity value. There is no separate representation of the row and column elements: row j and column j are both represented by a single stimulus point j. The solution configuration provided by CPA will consequently be a compromise between the upper and lower triangular elements, since the diagonal elements are ignored.

This program is particularly well adapted to the analysis of data collected by the method of conditional rank order, or where the user wishes to ignore asymmetries and obtain a single solution for the row and column elements. A typical example occurs for sociometric data, where each individual in a group is asked to provide a rank order of those closest or most similar to him or herself. Collected together, such data form a square, row-conditional set of rankings whose diagonal entries are irrelevant. For example, Gleason (1969) analyses an aggregated set of sociometric rankings obtained from the last four weeks of Newcomb's (1961) study of the process of friendship formation in an experimentally-monitored men's dormitory at the University of Michigan.

In the original analysis, Gleason analyses the matrix as two-mode data (cf. section 6.2.3) with the individual-as-row element (the giver of the friendship choice) being represented as one point and the (same) individual-as-column element (the receiver of the friendship choice) as a distinct point. Considerable attention is paid not only to the inter-personal distances but also to the 'intrapersonal separations' of the individuals' two 'ideal points'. As might be expected, the intra-personal distances are small compared to the interpersonal ones, but appear to be otherwise uninterpretable. Gleason (ibid. p. 118) coyly concludes, 'Unfortunately, very little else can be said about these distances at this time', and indeed he himself symmetrises the data for performing hierarchical clustering. Such considerations strongly indicate the use of a one-mode conditional proximity approach of MINICPA.

Whilst this example is more useful as an example of what can be done than providing a good illustration of its utility, further procedures for scaling such sociometric data are given in Breiger, Boorman and Arabie (1975), and MINICPA provides a useful addition to the methods discussed there. The Newcomb data form the test data for the MINICPA program in the MDS(X) version.

6.1.3   Triadic data analysis (TRISOSCAL)

Concisely: TRISOSCAL (TRIadic Similarities Ordinal SCALing) provides:

internal analysis of a set of triadic (dis)similarity measures
by a Minkowski distance model
using a local or global monotonic transformation of the data.

Triadic data are collected especially by psychologists, sociologists and anthropologists wishing to elicit the constructs which subjects use in making judgments of similarity. The basic idea is that the subject is asked to consider groups of three objects at a time, taken from the full set. Two forms are in common use:

partial triadic data, where from each presentation of the three objects the subject is asked to pick out only the single most similar pair; and

full triadic data where both the most similar pair and the least similar pair are picked out. The intermediate pair can then simply be inferred.

Although the method of triads is a useful technique for data collection, the number of triads increases very rapidly with the number of objects (for $p = 5, 10, 15$ and 20 the number of triads is 10, 120, 455 and 1140 respectively). Obviously, beyond about $p = 8$, the presentation of the full set of triads becomes totally unfeasible and very taxing on the subject. Burton and Nerlove (1976) give a full description and discussion of experimental designs for minimising the number of triads presented whilst maximising the information gained.

The most important advantage which triadic data possess compared to simpler forms is that contextual effects of judgment can be directly examined—by examining whether the similarity between two objects remains the same when the third element is changed.* It is therefore unfortunate when triadic data are turned into 'vote-count' data before scaling, since the effect is to obliterate the triadic information. (In brief, the vote-count method consists of counting the number of

---

*The assumption that the judgment remains unchanged in the presence of irrelevant alternatives is important in a number of theories of choice, preference and social welfare; see Rescher (1969).

times that object j is judged more similar than object k in the data.) Roskam (1970, p. 406) has shown that such a procedure badly misrepresents the order information in the data and often results in ill-fitting scaling solutions.

The TRISOSCAL program provides a direct method for scaling triadic data non-metrically by a distance model. It differs from Roskam's original MINITRI program in allowing the user to decide either that the order information implied across *all* the triads be fitted systematically ('global stress'), or only that the separate orders *within* each triad be fitted.

The distinction can be illustrated as follows. Suppose triadic data have been collected from a group of individuals. That being so, it is quite likely that when presented with objects (A, B, C) one subject will decide that (AB) is most similar and (AC) least similar (implying that $d(A, B) \leqslant d(B, C) \leqslant d(A, C)$). Another subject when presented with the same triad, may decide just the opposite—that the pair (AC) is most similar, and (AB) least similar (implying that $d(A, C) \leqslant d(B, C) \leqslant d(A, B)$). Both agree, by implication, that $d(B, C)$ is intermediate—but how shall the conflicting information concerning $d(A, C)$ and $d(A, B)$ be fitted? The answer proposed by Roskam's 'local stress' approach is: treat each subject's triad, that is, judgment, as a distinct entity, fit each of the three distances within a triad separately and then define 'the' overall fitting value as the average of the different disparity values. Hence there will be as many disparity or fitting values (d_{jk}^0) as occurrences of the pair (j, k) in the triads data. In this instance, there will be two distinct values of d_{ab}^0 and d_{ac}^0, but their respective arithmetic average (denoted $\bar{d}_{ac}^0$) will represent 'the' fitting value for the pair concerned. Roskam suggests that the form of raw stress to be minimised should be:

Triads: 'Local' Stress

$$\text{Stress}_0 = \sum n_{jk}(d_{jk} - \bar{d}_{jk}^0)^2$$

where n_{jk} is the number of times that the pair (jk) occurs in the set of triads to be analysed.

Roskam's approach obviously tolerates a good deal of inconsistent data, but is a sensible strategy when data from a set of individuals is combined, or where the number of objects is large.

Prentice (described in Coxon and Jones 1979, p. 49 et seq.) suggests a more restrictive and stringent approach: to require total consistency *between* triads and to count each and every infraction of transitivity in the stress value rather than averaging. Consider the following two pieces of triadic data:

Triad	Most similar (MS) pair	Least similar (LS) pair	Dissimilarity data implied
1 (A, B, C)	(AB)	(BC)	$(AB) < (AC) < (BC)$
2 (B, C, D)	(BC)	(CD)	$(BC) < (BD) < (CD)$

Taken together, the information from both triads is consistent (transitive) and implies the following order of data:

$$AB < AC < BC < BD < CD.$$

Prentice's global approach requires that the data be fitted by disparity values in the same order, and in particular he requires that the dissimilarity (BC) be fitted by the *same* disparity value in both Triad 1 and in Triad 2.

The two approaches and the different fitting values which result are illustrated in Table 6.2 for the set of data just given. Assume that there is a configuration of four points whose interpoint distances are as in the second column of the table. Column 1 gives the pairs in the two triads in their correct order according to the data, and column 2 gives the corresponding distances in the current configuration (note that allowance has to be made for two fitting values for (BC), although there can obviously be only one such distance in the configuration).

How well does the current configuration match the data according to the criteria of local and global stress? In the next three columns (3–5) the necessary fitting values are calculated using monotone regression.

Column 3   Local stress allows the fitting value for (BC) in the first triad to be different to the fitting value for (BC) in the second triad, so the d values are calculated separately within each triad, yielding $(BC)_1 = 2\frac{1}{2}$ (the block average of 3 and 2) and $(BC)_2 = 1\frac{1}{2}$ (the block average of 2 and 1).

Column 4   'The' fitting value for (BC) according to Roskam's local stress is defined as the average of its two appearances, i.e.

$$\bar{d}^0_{bc} = (d^0_{(bc_1)} + d^0_{(bc_2)})/2 = (2\frac{1}{2} + 1\frac{1}{2})/2 = 2$$

Column 5   The fitting values according to Prentice's global stress approach, which requires weak monotonicity over all the pairs, produce a *single* fitting value for (BC), which by block averaging over all but the first and last pair, gives a value

(1)	(2)	(3)	(4)	(5)	(6)	(7)
DATA Pair (ij)	DISTANCE d_{ij}	FITTING VALUE d^0 (local)	$\bar{d}^0$ (local)	DISPARITIES $\hat{d}$ (global)	RESIDUALS $(d - d^\circ)^2$ (local)	$(d - \hat{d})^2$ (global)
Triad 1 (AB)	1	1	1	1	0	0
(AC)	3	$2\frac{1}{2}$	$2\frac{1}{2}$	2	$\frac{1}{4}$	1
$(BC)_1$	} 2	{ $2\frac{1}{2}$	2	} 2	0	0
Triad 2 $(BC)_2$		$1\frac{1}{2}$	2		0	0
(BD)	1	$1\frac{1}{2}$	$1\frac{1}{2}$	2	$\frac{1}{4}$	1
(CD)	4	4	4	4	0	0
				Total	$\frac{1}{2}$	2

Raw Stress (local) $= \frac{1}{2}$
Raw Stress (global) $= 2$

Table 6.2   *Triadic data: illustration of local and global stress*

of 2. (It so happens that in this example both local and global approaches give the same fitting value for d_{bc}; this will not often be the case.)

The squared discrepancy values contributing to raw stress are given in column 6 for local and in column 7 for global stress, and the stress values are given at the foot of the table. Note that, as expected, requiring a complete ordered fit over all the pairs of both triads increases the badness-of-fit. Note also that global stress assumes the same value (2) for all but two pairs, but *is* weakly monotone with the order implied by the data. Local stress, by contrast, not only fits the same distance (*BC*) by two distinct values but also averages them to obtain the 'overall fitting value' which (as in this case) usually will *not* be in the same order as that implied by the triads:

$$\text{data: } AB < AC < BC < BD < CD$$
$$\bar{d}^0: \quad 1 \;\; < \;\; 2\tfrac{1}{2} \;\; \not< \;\; 2 \;\; \not< \;\; 1\tfrac{1}{2} \;\; < \;\; 4$$

In the case of inconsistent sets of triads, even greater inversions occur. When triadic information comes from a number of subjects or sources, highly inconsistent (intransitive) data often result and the user is faced with a difficult choice: either to choose global stress and risk technically degenerate solutions (since the inconsistencies can only be dealt with by imposing the same fitting value on a large number of pairs, capitalising on weak monotonicity) or to choose local stress, and lose all information about the order across triads as well as drowning same-pair inconsistencies by fitting averaged disparities. At least the Prentice approach signals the inconsistencies and global intransitivities by a high stress value—though at the disadvantage of increased computing costs.*

In such situations the user is advised to scale at least a random subset of data (to save computing time), using the global stress approach, and to compare resulting stress values and configurations with those obtained using local stress. Whichever stress option is chosen, triadic data are scaled in their integrity, and as far as possible consistency is kept within triads in the configuration. But only the global stress approach tries to represent information *between* the triads.

In some applications it is advantageous to insist upon between-triad comparisons by choosing the global approach, as when context effects are to be minimised. But by the same token context effects can best be studied by collecting triadic data and the user can then examine whether a given pair of objects tends to be fitted by approximately the same value. This can only be done by choosing the local stress approach.

6.1.4   The basic metric model (MRSCAL)
Concisely: MRSCAL (MetRic SCALing) provides:

internal analysis of two-way data of a lower triangle format of a (dis)similarity measure
by a Minkowski distance function,
using a linear and/or logarithmic transformation of the data.

---

*In Coxon and Jones 1978a, global stress_2 values of over 0.95 were frequently observed for such heterogeneous data sets, with corresponding local stress_1 values of around 0.20. In one case, 169 triadic comparisons of 13 occupations made by a set of policemen produced a global stress_2 value of 0.960 when scaled, and 75 out of the 78 global $\hat{d}$ values had the same value!

As we have seen, the assumption that data dissimilarities are a linear function of the distances of the solution historically precedes the monotonic assumption, and the earliest computational techniques for distance model scaling all assumed a linear (or ratio) transformation. Nowadays it is sensible to begin by scaling one's data by the non-metric model, thus making more defensible assumptions about the level of measurement of one's data. But since regular functions are special cases of the general monotonic family of transformations, the Shepard diagram obtained from non-metric scaling should be inspected to determine whether a more regular relationship is discernible between the data and the solution. If so, it makes eminent sense to go on to submit the data to a program such as MRSCAL, which implements the more regular linear and power transformations described in section 5.2.3. The examples in Chapter 3 provide illuminating illustrations of Shepard diagrams where a more regular relationship is evident. For the Scottish mileage data, the Shepard diagram (Figure 3.4) suggests an S-shaped or sigmoid power function*, although in the main range of distances (0.05 to 2.00) the relationship to the data is linear. In the small illustrative example of the similarity of eight crimes (Figure 3.2) the perfect strong monotone function is very close to being linear ($r = 0.97$). By contrast, in the case of the 'real data' example of scaling the co-occurrence frequency of occupational titles, the final Shepard diagram at iteration 23 (Figure 3.14b) shows a distinctly J-shaped (downwardly concave) form, again suggesting the use of a power function (in this case, a negatively accelerated exponential decay function).

As a matter of interest, the 2-dimensional configuration and its associated Shepard diagram obtained from a *linear* scaling of the seriousness of offences data are presented in Figures 6.1 and 6.2. Because the data are very well fit, the equation of the line relating the data to the solution distances is of interest, and attention focuses chiefly upon the slope, as in any other case of linear regression. The linear scaling transformation can be interpreted as indicating that, if the data are increased throughout by an additive constant of 0.251, then the distance between the points will on average be one quarter (0.241) of the difference between those saying they are 'unalike in their seriousness'. The advantage of a regular transformation function, then, is that it is possible to extrapolate beyond the original data (and in this sense 'predict' further data) if the model is correct. (In this particular example, we should not expect the transformation to be linear throughout its range, since there is an upper limit of 1.00 on the data values.)

The MRSCAL program in the MDS(X) series implements the linear and power models outlined in Roskam (1972) and contains options for implementing the city block and Euclidean distance metrics, among others.

6.1.5 Parametric mapping (PARAMAP)
Concisely: PARAMAP (PARAmetric MAPping) provides:

internal analysis of either a rectangular or a square symmetric two-way data matrix
by a distance model which maximises continuity or local monotonicity.

---

*In fact, a logistic function would best fit these data, but an ordinary logarithmic transform closely approximates its form.

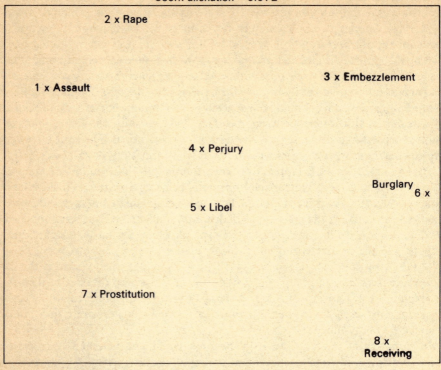

Figure 6.1   *MRSCAL 2-D solutions to data of Table 3.1*

The PARAMAP program accepts data either in the form of symmetric lower-triangular distances *or* in rectangular form consisting of profile values or spatial coordinates, but it only represents the objects (rows) in the latter case. The 'smoothness' or continuity transformation used in PARAMAP is the kappa family of continuity indices described in detail in 5.2.2.1 and in Appendix A5.1. The default values of the program parameters produce the simple 'normalised kappa' index (Appendix A5.1, equation 5), and the effect of these and other variants on the representation of the data are discussed in some detail at that point.

The main distinguishing characteristics of parametric mapping are:

(i)   the faithful preservation of the *local* information (small distances) around each point, virtually ignoring large distances (the user is thus given control of the degree of local monotonicity); and

(ii)   the 'flattening' of configurations into as small a dimensionality as possible, by increasing the size or variance of the largest distances.

The two are related: the price paid for preserving local information in a low dimensionality is the considerable distortion of global information. This fact should be borne in mind when interpreting PARAMAP solutions, since it contradicts the usual MDS maxim that configurations are globally stable, in the sense that the main features are reliably fixed but are locally unstable in that the points in a configuration can be moved around slightly without any major change in stress.

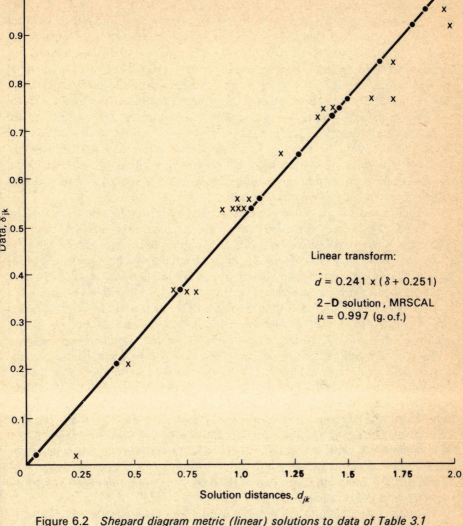

Figure 6.2   *Shepard diagram metric (linear) solutions to data of Table 3.1*

Moreover, because of the considerable non-linearity of the continuity rescaling transformation, any external properties the user wishes to represent should be mapped into a configuration using the *non-linear* option in PROFIT, which optimises an index akin to kappa. If hierarchical clustering is used to interpret a PARAMAP solution, only the initial stages of the clustering should be mapped into the configuration, because the largest distances are bound to be badly represented as a result of the continuity criterion.

The most dramatic examples of parametric mapping, which show these properties most obviously, occur for the mathematical structures of points defined at regular intervals on the circle (mapped into the line), on the sphere and on the torus, when mapped down into two dimensions (see Shepard and Carroll 1966, section 3.7). These form the test data for the MDS(X) program. The mapping of 20 points on a circle down onto a line is illustrated in Figure 6.3. In the original 2-dimensional space (the circle) the points are equally spaced; a continuity mapping attempts to preserve this property in the 1-space. But since the program is required to represent it in one dimension, this can best be done (producing only one major discontinuity) by tearing the circle at some point—indeed, at *any* one point. At the first iteration the adjacent points are equally spaced (except the two end points) but the ordering is entirely wrong—they are about four positions away from where they should be. By the 15th iteration the fit (κ) is nearly perfect and by the 30th the sequence of equally-spaced points is entirely correct except for the 'tear' in the circle between O and P. The Shepard diagram is not shown in this example (it forms a perfect quadratic function, as it should!) but the usual shape in the case of a PARAMAP solution is that of a fan (see Figures 5.3 and 6.4), with the larger dissimilarities fit by a wide range of distances (the broad end) and the smallest values fit extremely well (the narrow end).

Parametric mapping is defined and illustrated by a range of examples in Shepard and Carroll (1966), but the number of other applications in the published literature is disappointingly small. It is an especially useful procedure both for the analysis of two-mode profile data, when only the objects/stimuli are to be represented (ibid., section 2.1) and for mapping an already obtained configuration or set of distances into a yet lower-dimensional space. The former type of application is illustrated by Coxon and Jones (1979, pp. 140 et seq. and 1979, section U5.1), where subjects judged how appropriate 50 descriptions (such as 'A—would have a boring repetitive job') were of 20 occupations. Each occupation was then described by a 50-description profile, consisting of the percentage of subjects who considered that a given description was 'always' appropriate, and the resulting set of 20 profiles were then submitted to PARAMAP. Since PARAMAP is prone to produce local minima, and is also generally slow to converge, three different analyses were run, and the best yielded a kappa value of 1.09 in two dimensions. Figure 5.3 is the resulting Shepard diagram. The resulting configuration bore a good resemblance to the three-dimensional INDSCAL solution of similar data (see Coxon and Jones 1979, U5.2 for a discussion of the comparison), although only one dimension was identifiably the same—in effect, two of the INDSCAL solutions had been merged by the local monotonicity constraints.

The second type of application is illustrated by using PARAMAP to attempt to 'unbend' the highly non-linear horseshoe sequence of occupations described in 4.6 into a one-dimensional continuum. (Devotees of Conan Doyle will recognise that Holmes performed a similar feat, recalled in Kendall 1971a, p. 231.) The three-dimensional configuration was submitted to a one-dimensional parametric mapping, producing the instructive result illustrated in Figure 6.4.

It will be recalled from section 4.6 that graphical interpretation of a 3-D scaling configuration of occupational similarities yielded a 'horseshoe', bent at the end into the third dimension, representing skill (see Figure 4.5). How well can PARAMAP project this 3-space into what is believed to be a non-linear one-dimensional sequence? Using the default options for kappa, the result is given in Figure 6.4. The one-dimensional sequence does a good job in tracing out the horseshoe sequence, with the highly proximate occupations kept close along the line—as should be the

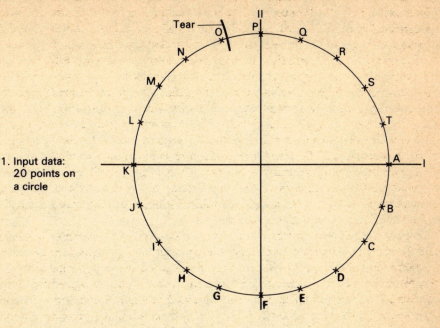

1. Input data:
   20 points on
   a circle

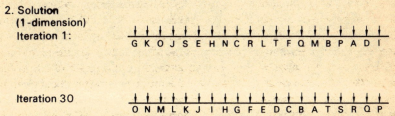

2. Solution
   (1-dimension)
   Iteration 1:

G K O J S E H N C R L T F Q M B P A D I

Iteration 30

O N M L K J I H G F E D C B A T S R Q P

3. Fit index

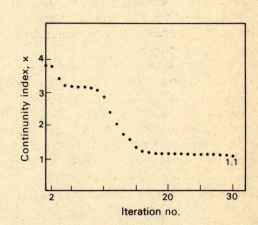

Figure 6.3   *PARAMAP maps a circle into a line*

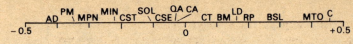

(a) PARAMAP 1–D mapping   (x = 1.07, with a = 1, b = 2, c = 1), iter = 30

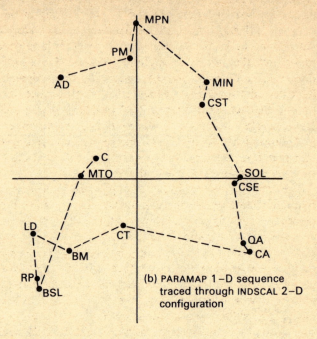

(b) PARAMAP 1–D sequence
traced through INDSCAL 2–D
configuration

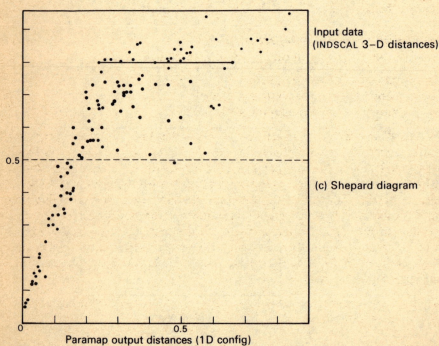

Input data
(INDSCAL 3–D distances)

(c) Shepard diagram

Paramap output distances (1D config)

Figure 6.4   *PARAMAP 1-D sequencing*

case with kappa giving so much weight to local monotonicity. When the PARAMAP sequence, but not the actual intervals, is mapped back into the original configuration (Figure 6.4), it can be seen that the 'cut' in the horseshoe sequence has been made between AD and C and not, as in the original interpretation, between BM and CA. Is the PARAMAP sequence wrong? Not really. The default values used make the solution virtually ignore long distances, so we should not expect that they will be faithfully reproduced. Look at the enormous variation in the PARAMAP distances of 0.5 and above for original configuration distances indicated above the line in Figure 6.4. For instance, input distances of 0.8 have output distances ranging from 0.24 to 0.66—almost half the total range! In fact a reasonable job is done in reproducing even the larger distances—after all, the ambulance driver-lorry driver distance, which is an important part of the original interpretation, is actually greater than the chartered accountant-commercial traveller interval which PARAMAP links.

The analysis followed in this example—preliminary scaling in multidimensional space, detection of non-linear sequences by graphical procedures and subsequent mapping a configuration down into a one-dimensional scale—may seem a cumbersome and complex procedure, but it is salutary to remember that the sequence would almost certainly *not* have been detected if we had proceeded immediately to a uni-dimensional scaling of the data. However, users should also be aware that the parametric mapping procedure assumes that the input data already represent distances in a space, and caution should be exercised before mistakingly submitting *any* dissimilarity data as input to PARAMAP. A useful initial step would be to make preliminary use of a program such as MVNDS (see 6.1.7) to produce a 'better behaved' set of distances from the original data.

6.1.6 *Hierarchical clustering schemes* (HICLUS)
Concisely: HICLUS (HIerarchical CLUStering) provides:

> internal analysis of two-way (dis)similarity data
> by means of a hierarchical clustering model,
> using a monotonic transformation of the data.

Hierarchical agglomerative clustering is an unusual inclusion in a set of programs for MDS, since it really belongs in the family of discontinuous clustering ('taxonomic') models for the analysis of dissimilarity data, and the solution consists of a discrete set of groups or classification,* rather than continuous spatial models of MDS. Nonetheless, HICLUS has so frequently been used in conjunction with MDS solutions that its inclusion as a utility is virtually mandatory.

In the language of cluster analysis, HICLUS includes two agglomerative procedures (that is, methods which begin by merging two most similar points into a single grouping and continue to merge points into successively more inclusive groupings). These two methods are:

single linkage (or nearest neighbour), referred to as the 'connectedness' or minimum method in HICLUS; and

---

*An elementary introduction to cluster analysis is given in Everitt (1974) and a more advanced mathematical treatment is contained in Jardine and Sibson (1971). See also 8.4.2.

complete linkage (or furthest neighbour or diameter method) referred to as the 'diameter' maximum method in HICLUS.

Since the two methods define 'the' distance between a cluster and another point as the minimum or the maximum intra-cluster distance respectively, they represent two extreme ways of representing the data and neither will be changed if the data are monotonically transformed. A set of data which can be perfectly represented by an HCS will obey not only the triangle inequality requirement of any distance measure (see A2.1.1), but also the more restrictive 'ultra-metric inequality', namely:

Ultra-metric Inequality
For all triples of distinct points (i, j, k),
$d(i, k) \leqslant \max \{d(i, j), d(j, k)\}$

In the case of perfect data, both the 'connectedness' and the 'diameter' methods give rise to the same hierarchical clustering. The HICLUS procedure and its applications in MDS were described in detail in 4.3.3.1.

6.1.7 Maximum variance non-dimensional scaling (MVNDS)

Concisely: MVNDS (Maximum Variance Non-Dimensional Scaling) provides:

internal analysis of two-way data in a lower-triangle format of a (dis)similarity measure

by a simple distance model

using a locally monotonic and variance maximising transformation of the data guaranteed to satisfy the triangle inequality criterion.

This program and its implementation are described in detail in Cunningham and Shepard (1974). MVNDS is like all the other programs in the MDS(X) series in that it rescales the data (dis)similarities into a set of corresponding distances. It differs from them in *not* producing a configuration of points in a continuous space to which the distances correspond. Indeed, the solution distances may well not be capable of being represented spatially at all. All the MVNDS scaling procedure requires is that the solution distances have two properties. They must

(i)   satisfy the metric axioms, especially the triangle inequality; and
(ii) be as close as possible to being a monotonic function of the data.

The basic distance model rescales the data into a set of disparities which are *as close as possible* to being distances. In any imperfect solution it is very likely that some triples of the disparities will *not* satisfy the triangle inequality, and this is one reason why they are sometimes termed pseudo-distances. In MVNDS the scaling problem is dealt with in a different way—by severing the link to a Euclidean (or other similar) space and requiring that the actual fitting values (the rescaled data) satisfy only the triangle inequality.

As they stand, these two requirements are not sufficient to obtain a (non-trivial) solution but need to be supplemented by a third, which serves the same purpose as seeking a solution in as low a dimensionality as possible in SSA:

(iii)   to maximise the variance of the distances.*

The MVNDS program maximises a goodness-of-fit index between the data and solution distances which has three component weights of the form:

$$\text{Index} = \text{W1} - \text{W2} - \text{W3}.$$

W1 represents the variance of the distances, W2 represents departure from weak monotonicity (raw stress) and W3 represents violations among the distances of the triangle inequality. (The default values give W3 thirty times the weight of W2 and one hundred times that of W1.) As the number of iterations increases, violations of the triangle inequality dramatically decrease, usually producing a set of rescaled quantities which perfectly satisfy the triangle inequality and other distance measure axioms.

From simulation studies it turns out that MVNDS is very well adapted to recovering distances derived from non-spatial structures (such as the length of paths through a graph or tree) to recovering non-Euclidean distances and to recovering distances which have been subject to a wide range of distorting transformations. (Incidentally, it does far better in this than the basic non-metric MDS model.)

The claims of the MVNDS model thus appear to be well-founded: MVNDS does a very good job at recovering a wide class of distances very accurately and is an excellent choice when the user is primarily interested in rescaling data rather than in producing a spatial configuration to mirror the data. A good deal of evidence suggests that many cognitive and semantic structures (networks, generative structures, associative graphs) and many aspects of organisations, life-histories, trade-flows, social networks etc., are better represented by non-continuous structures than by spatial models. Use of MVNDS as a preliminary step either to spatial or non-spatial representation is therefore a sensible choice in these applications.

An interesting application of MVNDS is the analysis of Rothkopf's experimental data based upon judgments of perceived similarity between 36 morse code (dot and dash) signals. He presented the 600 subjects with pairs of auditory signals (separated by a 1.4-second interval, with a gap of 3 seconds between the pairs) and asked them to say whether each pair was 'the same' or 'different'. These observations were converted into a matrix of confusion measures whose entries δ_{jk} gave the proportion of subjects saying that signal j and signal k were the 'same'. The data are presented in Shepard (1963) together with a non-metric MDS two-dimensional solution, readily interpretable in terms of (1) the number of dots and dashes, and (2) the predominance of dots over dashes. A subset of these data, restricted to 21 of the 36 signals, was re-analysed in Cunningham and Shepard (1974, p. 354 et seq.) according to the MVNDS model, with weights which ensured

*This corresponds to forcing a solution down into a lower dimensionality whilst still preserving the monotonic constraints of the data (cf. Shepard 1962). Intuitively, it helps to consider a triangle with unequal sides. Keeping the order of the distances, the variance will increase as the longest side is lengthened until it equals the sum of the other two sides—thus lying down in a one-space. In general 'when variance is forced upward by ... stretching large distances and shrinking small distances, the dimensionality of the least space that will adequately fit a set of points tends quite generally to decrease' (Cunningham and Shepard 1974, p. 341).

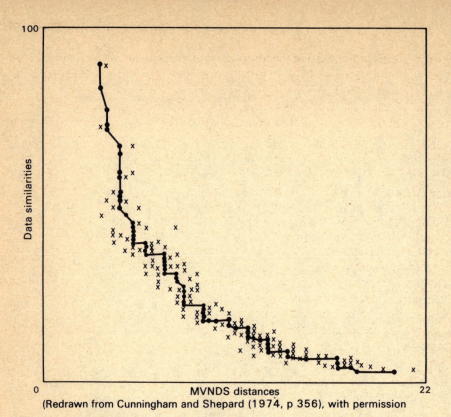

(Redrawn from Cunningham and Shepard (1974, p 356), with permission

Figure 6.5   *MVNDS solution to Rothkopf's morse-coded confusion data*

that the solution distances conformed to the triangle inequality, but which did not emphasise conformity to monotonicity very strongly, since these were known to be errorful data. The resulting Shepard diagram is presented in Figure 6.5. (A smaller subset of these data provide the test data for the MVNDS program in the MDS(X) series.) It is instructive to note that the monotone function is very much like a negative exponential 'decay' curve—a relationship expected on theoretical grounds and frequently found for such data (Shepard 1980, p. 391). It also turns out that the MVNDS 'distances' are in fact very close to Euclidean distances of the 2-dimensional non-metric MDS solution (correlating at 0.954).

When MVNDS is used with empirical data, it is advisable to set W3 at a high level to ensure conformity to the metric axioms, and then seek a compromise between the maximum variance (W1) and monotonicity (W2) conditions, with W2 set in the vicinity of 2.0 (see Cunningham and Shepard 1974, pp. 357–8 for discussion of suitable strategies). In any event, users are strongly recommended to obtain more than one solution, using different values of W2.

6.1.8 *Simple conjoint composition models* (UNICON)
Concisely: UNICON (UNIdimensional CONjoint Measurement) provides:

internal analysis of a table of values representing the composite or conjoint effect of up to five variables (or facets) on a single dependent variable
by either additive, subtractive, multiplicative (and more complex) models, using a monotonic transformation of the data.

The additive sub-model of UNICON is the one most commonly used and will form the basis of presentation here. In brief, the purpose is to find *uni-dimensional* scale values, or 'utilities', for each category of a table of two or more nominal variables ('ways' or 'facets' in Guttman's terminology) which, when combined by addition, best reproduce the rank order of the entry in the table.*

UNICON is most useful when the researcher's primary concern is to estimate the quantitative effect which each of a number of 'events' (nominal variable categories) has upon a dependent variable, and/or investigate in what way such effects combine to produce the dependent variable value. This model has a lot in common with a number of currently popular and frequently-used methods for analysis of two-way (and higher) tables of data, whose entries may be values (mean, median) of a quantitative dependent variable (e.g. the expenditure on food in households where one parent is from class x and the other from class y) or else a percentage frequency, as in the usual contingency table. Such methods include conventional analysis of variance (ANOVA) and Tukey's related, but more robust and resistant, 'median polishing' technique (Mosteller and Tukey 1977, p. 186 et seq.)† in the first case, and log-linear analysis of contingency tables in the case of percentaged data.

ADDIT can perform both types of analysis, but seeks the less restrictive *ordinal* rescaling of the table entries. It is therefore especially useful when the researcher has used a rating scale method to collect data and/or wishes to make rather modest measurement assumptions about her data. Note especially that UNICON solution scale values are uni-dimensional (one uni-dimensional scale for each way of the data) ‡ and are only interval-level values, i.e. unique up to a linear transformation. In both these respects UNICON differs from other MDS programs.

Very commonly, when two-way data are analysed by conventional ANOVA, the additive model has to be extended to include an 'interaction term' to allow a separate effect for each entry in the table over and above the purely additive effect of each row and column. In one sense, the interaction effect is evidence that the simple additive model does not hold, and statisticians have frequently looked for a transformation of the data which will reduce or remove the interaction effects. Since considerable interaction effects are frequently found in empirical research (and especially in social psychological studies based upon rating data whose measurement status is obscure) it is always worth at least entertaining the hypothesis that interaction effects observed from conventional ANOVA may in fact arise from the somewhat arbitrary measurement scale of one's data. In at least one instance, ordinal additive analysis can be shown to render data perfectly additive when the authors initially claimed the reverse. Sidowski and Anderson (1967) had subjects rate the attractiveness of working in certain professions in cities of varying sorts (e.g. *doctor* in a city chosen to be *moderately high* in attractiveness as a place

---

*Following Roskam's convention this additive model and its iterative implementation is referred to as ADDIT, which was the name given the program in an earlier release of MDS(X). The measurement theory underlying the model (conjoint measurement) is discussed in detail in Krantz et al. (1971) and a simple overview is presented in Coombs et al. (1970, pp. 25–30).

†See Coxon and Jones (1978b, pp. 66–86) for an example of the use of median polishing in conjunction with MDS scaling procedures.

‡Strictly, this makes UNICON a method for analysing multi-mode, multi-way data, but in practice several ways refer to the same mode.

to live) and a conventional ANOVA yielded a highly significant interaction term. Krantz et al. (1977, pp. 445–7) show that an ordinal rescaling of the data yields a perfect additive fit and that:

> the interaction between city and occupation, therefore, is attributable to the nature of the rating scale, because it can be eliminated by appropriate rescaling.
>
> (ibid. p. 446)

This conclusion has been challenged in Birnbaum, 1974.

The algorithm used in ADDIT and UNICON is described in the MDS(X) documentation and the additive case is discussed in some detail in Roskam (1968, ch. VI) and Kruskal (1965). Basically, the initial scale values are provided by pseudo-random numbers, and Kruskal's weak monotone regression used to obtain fitting values. Stress$_2$ is then used to estimate fit, and the values are improved by a gradient procedure similar to that used in the basic non-metric model (see Kruskal 1965, pp. 261–2). Users should be on their guard against degenerate (artificially low stress) solutions which capitalise upon the *weak* monotonicity of Kruskal's procedure. The principal means of detecting such degeneracy is to compare the number of distinct values (equivalence classes) in the data, in the disparities and in the solution values. If there are significantly fewer distinct values in the disparities it is advisable to check where these occur, and especially whether they are concentrated in any particular category of the table variables. In any event, it is probably sensible to err on the side of caution if ties occur in the data and use the secondary approach to ties.

An example of the use of the ADDIT sub-model of UNICON is given by the data on mean fertility of 1,893 married couples in a national (England and Wales) sample in 1949, originally collected by Berent (1952) and subsequently analysed by Blau and Duncan (1967) and Hope (1972). Average fertility is assumed to be the additive effect of the typical fertility behaviour of (i) the social origin class of the husband, and (ii) of his current status. In the first version, the class fertility effects are assumed to be different, so the model takes the form

$$x_{ij} = \mu + a_i + b_j$$

where μ is the overall mean fertility, a_i is the effect of the row (social origin) class category i, b_j is the (possibly different) effect of the column (present status) class category j, and x_{ij} is the predicted mean fertility of couples whose husbands came from class i and are presently in class j. (This model is coded in UNICON simply as: MODEL $A + B$.) If an additive model holds well and there are no significant interaction effects it can be asserted that the couple's fertility is simply attributable to the influence of the husband's (though why consider only the husband's?) origin and destination status and is not separately affected by distinct social mobility effects (represented by particular combinations, or moves between, class categories). Obviously, the analysis of residuals, here representing the interactions, will be central to the analysis.

The second variant (often referred to as the 'halfway hypothesis') simply assumes that the fertility effect of the classes is the same, whether they feature as the origin or the destination statuses. (In terms of UNICON, this simply constrains the a_i and b_i effects to be the same, and is simply coded as: MODEL $A + A$). In this

case, unlike the first hypothesis, the effects will be symmetric—the predicted fertility will be the same whether the husband has moved from i to j or from j to i.
Using conventional ANOVA, HOPE (1972, p. 98) concludes:

> Both analyses result in the rejection of the halfway hypothesis and in the acceptance of an alternative which involves an effect of direction of social mobility on fertility, the downwardly mobile being more fertile than the upwardly mobile, with the non-movers in the middle.

If the fertility data are assumed to be ordinal (hence leading to a more resistant analysis where more extreme values have relatively little effect), do the same conclusions emerge?

The scale values for the two models and the tables of residuals are given in Table 6.3. The scale values are given in the row and column stubs, and the entries within the table give the residuals $(d - \hat{d})$ from the monotone additive model.

I   MODEL: SAME EFFECTS
('Halfway' or symmetric mobility effects on fertility)
(MODEL: $A + B$, with $A = B$ Stress$_2$ $(\hat{d}) = 0.263$)

SOCIAL CLASS OF:	DESTINATION				Row Effects
	I	II	III	IV	a_i
ORIGIN					
I	0	−0.857	0	−0.141	−0.252
II	0	+0.138	+0.589	0	1.153
III	0	0	0	0	1.194
IV	+0.267	0	−0.542	+0.541	2.277
Column Effects: b_j	−0.252	1.153	1.194	2.277	

II   MODEL: DIFFERENT EFFECTS
(Asymmetric mobility effects on fertility)
(MODEL: $A + B$, with $A \neq B$: Stress$_2$ $(\hat{d}) = 0.160$)

	DESTINATION				Row Effects
	I	II	III	IV	a_i
ORIGIN					
I	0	−0.140	+0.088	0	0.504
II	0	0	+0.140	−0.165	0.814
III	−0.084	+0.028	+0.008	+0.069	1.355
IV	−0.003	+0.118	−0.219	+0.150	1.435
Column Effects: b_j	−0.141	0.911	0.881	1.250	

Table 6.3   UNICON *(monotone additive) analysis of Berent's fertility data* (Solution scale values are given in the row and column stubs. Entries in the tables are the residuals $(d - \hat{d})$ from the monotone additive model.)

Compared to Hope's linear analysis of the same data (see especially Hope 1972, pp. 88 and 98–9), the conclusions from the monotone analysis presented here are markedly different. The order of the scale values is the same, as we should expect, and the linear correlation of scale values from the linear and monotone analyses is quite high ($r = 0.90$), but analysis of the residuals shows a rather different structure. In particular, there is no consistent pattern in the sign of the residual values according to whether the subjects are upwardly or downwardly mobile (Hope expects positive residuals for the latter and negative values for the former) and the sources of significant interaction are different.

In this example the differences between a linear and monotone analysis of such data should not be taken too seriously—after all, a linear analysis makes obvious sense for a variable such as the average number of children. But when the dependent variable is based on ratings data, as is usually the case, the effects due to the variable and the purely artefactual effects due to the arbitrary nature of the rating scale are likely to be confounded. In such instances it is eminently sensible to rely primarily on a monotonic analysis which makes no assumptions about the interval nature of the data.

In addition to ADDIT analysis, UNICON provides for subtractive models (of the form: $X = A - B$) and multiplicative models (of the form: $X = A \times B$), and more complex mixed models, such as: $X = (A \times B) + C$.

6.2 Two-Mode Data

In two-mode data the input data matrix represents *two* (usually distinct) sets of objects, each of which is represented separately in the solution. The data matrix is normally rectangular (since the number of row and column elements is usually different), asymmetric, and row-conditional, in the sense that each row of the data is fitted separately and data comparison between rows is considered illegitimate.

In the case of simple composition models, the UNICON program discussed in the previous section is the appropriate model.

In other cases, (vector or distance models) a further distinction must be made between whether the analysis is internal or external (see 5.1.2). In most applications, the column elements represent the objects or stimuli and the row elements represent individual sources of data (often subjects), and the number of columns is much less than the number of rows. Consequently the positioning of the objects will depend very heavily upon satisfying the constraints of as many individuals' data as possible according to the assumptions of the model. Therefore great care should be taken to interpret the stimulus configuration with explicit reference to the location of subject points (or vectors). This matter is taken up below.

By contrast, in an *external* analysis, one part of the configuration—usually that part representing the objects (columns)—is provided by the user and (usually) remains fixed during the analysis. The data are then used to locate just the subject points within the stimulus configuration.

6.2.1 *External mapping by the vector model* (PREFMAP IV *and* PROFIT)
Concisely: PREFMAP (PREFerence MAPping) (Phase IV) provides:

external analysis of two-way, row-conditional data
by a scalar products model,
using either a monotonic or a linear transformation of the data.

Concisely: PROFIT (PROperty FITting) provides:

external analysis of a configuration
by a set of property ratings or rankings in row-conditional format
by a scalar products model,
using either a linear or continuity transformation of the data.

Given

(i)   a fixed configuration of p stimulus points in a specific number of dimensions,
and
(ii)   a rectangular data matrix of N rows (one for each subject or property) and
p columns (identical to the configuration points),

these two programs map each of the N subjects/properties into the stimulus configuration as a vector, pointing in the direction in which the data values—preferences, property values, similarities—are increasing. For convenience, all the vectors pass through the centroid (centre of gravity) of the stimulus configuration. (Vector representation was discussed earlier in 4.4.1.)

What differentiates the two programs is simply the rescaling transformation option which can be requested:

(1)   PREFMAP-IV
The rescaling transformation can either be *linear* or *monotonic* (strictly speaking, quasi-monotonic, meaning that the procedure first performs a linear transformation and then goes on to a set of iterations using monotone regression). If the monotonic option is chosen, three further options exist, depending on whether ties exist in a subject's row or data:

(a)   no ties exist (FIT(1)).
(b)   ties are treated as equalities (secondary approach, FIT(2)).
(c)   ties are treated as indeterminate (primary approach, FIT(3)).

(2)   PROFIT
The rescaling transformation can be either *linear* or a *simple continuity* (kappa) function (see 5.2.2.1). In the former case, linear PROFIT is formally equivalent to PREFMAP-IV with linear fitting, although the algorithm used to produce the solution differs somewhat.

External mapping by a vector model is extensively used as a means of interpreting configurations and identifying dimensions; these uses were discussed in section 4.4.1 at some length.

External mapping is also often used to map subjects' preferences into an already obtained stimulus space. There is much to be said for analysing preference data in this way, since the evaluation of things may often be quite distinct from the cognition or adjudged similarity of things. We may agree entirely on the characteristics and relative similarity of a set of regimes, people, foods, books, but disagree violently on their merits or over which we prefer.

A full discussion of such analyses and a number of examples of applications of this sort is provided in Carroll (1972, especially pp. 130–46).

External mapping is also particularly useful when the researcher wishes to represent information within an already known or physical configuration—such as a geographical map or the plan of a machine assembly. In this way, the directions can be estimated in which a particular plant species increases in a botanical area, or social deprivation increases in a city, or strains increase within a machine, or travel priorities lie in a country, can be readily assessed. It is also possible, using the other distance sub-models of PREFMAP, to see whether these data would be better represented as a point—an 'ideal point'—and if so, what sort of more complex distance might best fit the data.

The significant information which should be attended to in an external vector analysis is, first, how well a particular row of data (subject, property) fits in the configuration. (The goodness of fit depends, of course, on the transformation chosen: a linear relation is more restrictive than a monotonic one and therefore a vector of data is bound to fit better—or at least no worse—monotonically than linearly.) Given acceptable fit, the most important information conveyed by a vector is simply its direction (in none of these programs is the length of the vector relevant) and in comparing vectors, the angle of separation is crucial and represents the correlation between the subjects' data. Communality or concentration in a set of vectors, akin to the cluster in distance representations, will show as a tightly bound sheaf, and divergence or difference is signalled by empty sectors.

A brief illustration of the results of using PREFMAP-IV for the analysis of the occupational data described above is given in Figure 6.6. The ratings of (a) social usefulness and (b) earnings averaged within nine sub-groups of subjects were mapped into the configuration (see Figure 4.6) obtained from scaling the similarities data. (In each case, the analysis of variance in directions of the vectors gives a significant difference between the groups of less than one per cent: see Table 4.16 and 4.20 in Coxon and Jones 1978a.) Note, first, that the vectors are heavily concentrated, in each case, within a small sector. The average 'social usefulness' vector is located in a NNE direction (at 62°) and the variability in the groups' vectors is contained within 34°; nine-tenths of the unit circle is empty. In this direction, the people-oriented, but low paid occupations ('vocations') are located, followed by less educated/lower paid and more educated/better paid through to the least socially useful. Notice the high degree of agreement between disparate groups—student teachers, engineers and clergy agreeing almost entirely (at least insofar as group averages go). By contrast, the average 'earnings' vector orders the occupations in an entirely independent manner (separated by just over 90°, representing a correlation of virtually zero) in a SE direction, but again with little spread, the extreme vectors subtending an angle of merely 30°. However, in this instance there is a detectable difference between the professionals' (more accurate) estimate and the working class groups' estimates, which tend to underestimate the professional occupations' remuneration compared to their own. It should be emphasised that present data are averages: the same conclusions do not necessarily hold when examining individual data.

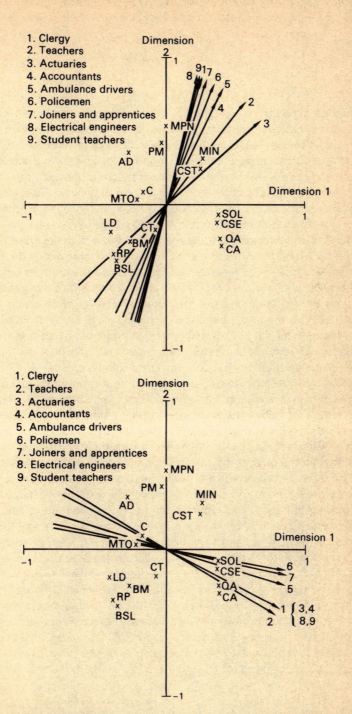

1. Clergy
2. Teachers
3. Actuaries
4. Accountants
5. Ambulance drivers
6. Policemen
7. Joiners and apprentices
8. Electrical engineers
9. Student teachers

Figure 6.6   *PREFMAP-IV: external mapping of two criteria into INDSCAL configuration*

6.2.2   *Internal mapping by the point-vector model* (MDPREF)

Concisely: MDPREF (MultiDimensional PREFerence Scaling) provides:

internal analysis of two-way preference data in the form either of a row-conditional matrix or of a set of paired comparisons matrices
by a scalar products (point-vector) model,
using a linear transformation of the data.

Note that MDPREF is an *internal* form of analysis, positioning stimuli points and subject vectors simultaneously from the data, and is a *linear* procedure: data are assumed to be at the interval level of measurement. As a form of two-mode factor analysis, MDPREF is becoming increasingly popular for analysing preference data, personal constructs rankings* and semantic differential ratings data. Since the solution is analytic rather than iterative, it is a computationally cheap and efficient procedure and the results are often a good deal more stable than for other internal two-mode models such as multidimensional unfolding (MINIRSA, q.v.).

From the user's point of view the main difference between the external (PREFMAP-IV) and internal (MDPREF) vector models is in the way of interpreting the solutions. In the external case, it will be recalled, subjects are located within a *fixed* reference configuration and the location of subjects' vectors could with some confidence be referred to or interpreted in terms of the stimulus locations. For internal analysis this is not true: the stimulus points are located in such a way that as many as possible of the subjects' data are fit well and the stimulus configuration can only be 'read' by direct reference to the location of the subject vectors.

The MDPREF algorithm is described in the MDS(X) documentation and in Carroll (1972, pp. 123–9). Basically, the algorithm forms the major and minor product moment matrices from the original rectangular data matrix (called the 'first score matrix' in the MDPREF terminology) and obtains the latent roots of those matrices. (These give a good estimate of the 'true'—or at least the lowest acceptable—dimensionality of the data.) The location of the stimulus points and subject vectors are then found by producing a factoring or decomposition which gives a 'second score' matrix which best fits the data in the number of dimensions chosen by the user. The model has already been discussed above in 5.3.2.

Issues in interpretation and application of the MDPREF program are best illustrated by reference to what is now a quite well-known data set: the Bollen-Delbeke data on family size and composition preferences. It will also be used to illustrate the corresponding distance model analysis (MINIRSA).

In 1960 Bollen collected data from psychology students at the Catholic University of Louvain on their preferences for families, which differed in terms of the number and sex of the children. In all, 21 such stimuli, (family size compositions) were defined—all possible compositions from no children up to families of size five. These are illustrated in Figure 6.7. Each subject was then given each of the 210 pairs of stimuli, e.g. (3 sons, 2 daughters) *vs* (1 son, 1 daughter), and asked which he or she preferred. The data for 80 subjects (40 male, 40 female)

---

*In many ways MDPREF resembles (and is indeed superior to) the INGRID program developed by Slater (1960) for the analysis of repertory grid data, frequently used in personal constructs analysis in psychology and sociology. See Tagg (1979).

formed the basis for subsequent analysis, and were analysed by Bollen and later by Delbeke (1968) and Coxon (1974). The data exist both in the original form as a set of 80 pair-comparison dominance (0, 1) matrices and as a set of 80 preference

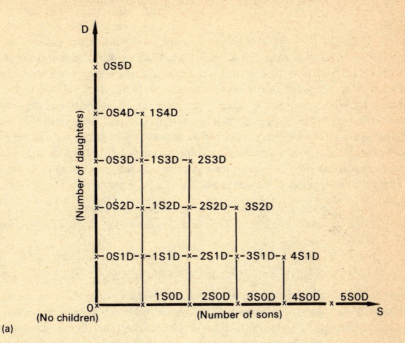

(a)

Summary information on rank scores for different family size compositions*

Stimuli	Code	Range	Mean	Variance
0 children		0–4	0.2	0.4
1 son	ISOD	0–12	3.6	7.3
2 sons	2SOD	2–17	8.1	13.1
3 sons	3SOD	5–16	9.7	6.6
4 sons	4SOD	4–18	10.0	9.6
5 sons	5SOD	0–20	8.7	20.6
1 daughter	0S1D	1–11	2.6	4.4
2 daughters	0S2D	2–13	5.6	7.7
3 daughters	0S3D	2–12	6.2	5.0
4 daughters	0S4D	1–13	5.8	5.9
5 daughters	0S5D	0–13	4.8	11.8
1 son, 1 daughter	1S1D	4–19	11.0	17.5
2 sons, 1 daughter	2S1D	10–20	15.1	6.1
3 sons, 1 daughter	3S1D	10–19	15.9	3.5
4 sons, 1 daughter	4S1D	5–20	14.8	12.6
1 son, 2 daughters	1S2D	6–20	12.5	7.5
2 sons, 2 daughters	2S2D	13–20	17.9	2.3
3 sons, 2 daughters	3S2D	12–20	18.4	4.6
1 son, 3 daughters	1S3D	2–18	12.0	8.7
2 sons, 3 daughters	2S3D	9–20	17.1	5.6
1 son, 4 daughters	1S4D	1–18	10.1	17.5

(b)   *Highest preference is a rank of 20, and lowest preference has a rank of 0.

Figure 6.7   *Delbeke family size and composition data: summary*

rankings, formed by summarising the rows of each dominance matrix to produce a 'vote count' preference order, since the subjects were remarkably consistent.* A basic summary information on the rankings data is provided in Figure 6.7b. Several points should be noted. First, the 'no children' stimulus had so little variation (virtually everyone preferred it least) that it was removed from subsequent analysis. Its inclusion led to the structure in scaling solutions being distorted, and the universal dislike of no children meant that its location was highly unstable, varying from run to run—being located at any position so long as it was maximally distant from ideal points in the distance model, and maximally opposite in direction to the preferred regions in the vector case.

Secondly, there is marked preference for large, mixed families, with the composition of 3 sons, 2 daughters having highest overall preference. The data, it should be remembered, were collected from unmarried Catholic students before the changes in attitude to and practice of birth control following Vatican II. Other characteristics are also evident, and are recognisable in the scaling solutions in differing ways, depending on the model.

(i)   For each given single-sex family size, all-boy families are preferred to all-girl families, and the difference in average preference of boys increases systematically with the overall size of family.

(ii)   For every family size, a mixed-sex composition is preferred.

(iii)   A preponderance of boys is preferred in mixed-composition families.

MDPREF was applied to both the preference scores and the pair-comparison data. Solutions were sought in two and three dimensions, for men and women subjects separately, though only that for the male subjects is reported here. The preference score and pair comparisons data produce almost identical solutions in each case. Inspection of the roots of the first-score matrices strongly suggests that a two-dimensional solution is adequate—the percentage of variation accounted for by the first four dimensions is 75 per cent, 14 per cent, 6 per cent, 1 per cent, so 89 per cent is concentrated in the first two dimensions and little is gained by adding any subsequent dimensions. The two-dimensional configuration for males is presented in Figure 6.8. (The female data form the test data for the MDPREF program in the MDS(X) version.)

First, let us concentrate on the location of subject vectors, recalling that for a two-dimensional MDPREF solution their termini (end points) are normalised to unit length and will therefore lie on a circle. However, virtually all subjects' vectors occupy just over one-quarter of the circle, indicating a quite high degree of consensus in their preferences. In internal analysis the position of subject vectors should take priority in interpreting the joint space. It is often helpful to begin by locating an average subject and reading back along from the vector end, through the origin of the space to the other side of the circle, noting how the stimuli project onto it. (The location of the dimensions is of course arbitrary in a vector model so they do not need interpretation, but the origin is significant as the centroid of the stimulus points and the point through which all subject vectors pass.) In this case

---

*See Coxon (1974, p. 197, ftn 11) for details of tests of consistency. The average coefficient of consistence was 0.94.

JOINT SPACE

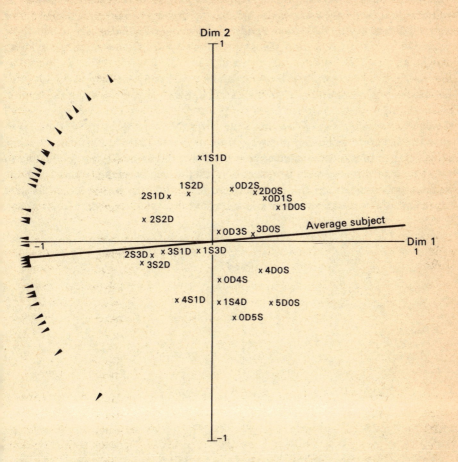

Figure 6.8 *MDPREF analysis of family size and composition data 2-D, males only*

the most popular stimulus (3S2D) is the first, most preferred one, followed closely by other fairly large, mixed compositions, then into larger single-sex families, and finishing with the least preferred single son and single daughter compositions. (The projections mirror fairly accurately the mean ranks given in Figure 6.7b.)

Secondly, what do the differences in vectors signify? The subject at one extreme (located in the NNW direction) clearly prefers much smaller families and is not by any means as concerned with the balance between sexes. By contrast, the subject at the other extreme (in the SSW direction) is greatly in favour of very large family size whether mixed in composition or not. On the whole, virtually everyone prefers mixed to unmixed composition, and the greatest variation is on the size of the family composition.

The structure of the stimulus configuration, then, should be interpreted by reference to the subjects' vectors in internal analysis and it is important to realise that the 'natural' lattice-like structure of the stimuli (see Figure 6.7a) cannot be discerned in the stimulus configuration. (Try connecting the sons' axis, denoted on Figure 6.8 as OD1S, OD2S, OD3S, OD4S, OD5S, and the daughters' axis and you will note that they follow a roughly parallel and non-linear sequence, with a major shift in direction at OD2S and 2DOS. In no way can the lattice structure be discerned in this configuration.) Nonetheless, a very coherent structure *is* evident, which happens to be part of a radex (see 4.5), and this is illustrated in Figure 6.9. A semi-circle can be drawn which divides the *mixed* from the single sex or *unmixed* family compositions, and a set of lines can be drawn emanating from the 'centre' which divide the space into sectors corresponding exactly in this case to the overall *family size*. Another way of describing the radex (Shepard 1978, p. 57 et seq.) is to view the structure as having polar co-ordinates, with latitude (distance from the centre of the radex located in an approximate way in Figure 6.9) corresponding to the degree of mixedness, and incidentally to the degree of preference, and the angular position around the perimeter corresponding to the size of the family. Moreover, the three characteristics noted above from the initial data analysis can

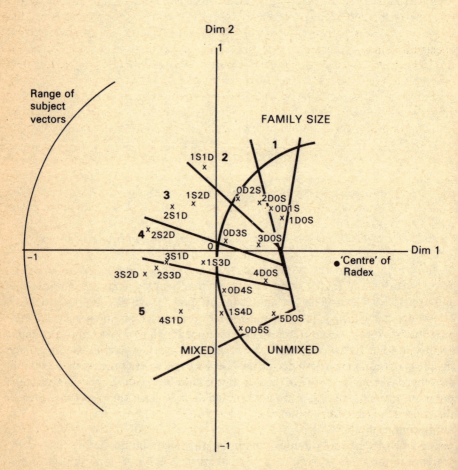

Figure 6.9   *Radex structure of family preference MDPREF solution*

in most cases also be 'read out' in a manner consistent with the form of the vector model:

(i) *All-boy families are preferred to all-girl families, for given single-sex family-sizes.* Concentrating on single-sex pairs of fixed family-size, the points representing all-boy families are systematically to the left of (more preferred than) all-girl families of the same size.

(ii) *A mixed sex-composition is preferred with a given family-size.* Within each given family-size the points representing mixed families are consistently to the left of (preferred to) those representing unmixed family composition.

(iii) However, the inference that *within* a given type of mixed composition family, a preponderance of boys is preferred does not seem to be detectable.

(The same characteristics will be recognisable, but in a different and apparently unrelated representation, when the distance (unfolding) MDS solution is given later in section 6.2.3.)

A final caution in the use of MDPREF. The program allows users to remove either the row effects (individual response-style) and/or the column effects (removing the 'mean utility' or consensus) of the first-score data matrix. The removal of row effects rarely has important consequences, but the removal of column effects will produce a much wider dispersal of subject vectors, and individual differences then become the major focus of the analysis (see Heiser and de Leeuw 1979, p. 28 et seq.). On the other hand, *double*-centring—removing both column and row effects—actually turns the vector model into a distance model and typically leads to an over-estimation of the appropriate dimensionality (see Carroll, 1970, p. 278). It should be used with considerable caution, if at all.

6.2.3 Internal mapping by the distance model (MINIRSA)

Concisely: Multidimensional Unfolding provides: MINIRSA (Rectangular Space Analysis) or

> internal analysis of two-way data in a row-conditional format of a (dis)similarity measure
> by a Euclidean distance model
> using a monotonic transformation of the data.

The basic multidimensional point-point distance ('unfolding') model was described in 5.3.3.1, and the algorithm is described in the MDS(X) documentation and in Roskam (1979, pp. 300–4). MINIRSA positions each subject *as a point* (the 'ideal point' or single point of maximum preference) in a joint space with the stimuli, also represented as points, such that the rank order of the distances from the ideal point to each of the stimuli is as close as possible to being in the same rank order as the subject's preferences (or other similar) data.

In the current MDS(X) series a metric variant of multidimensional unfolding is possible using the 'quasi-internal' options* of PREFMAP-III.

*INIT (0), FIT (0) and S-PHASE (3). See 7.5.5.2.

The *non-metric* internal analysis of rectangular data by multidimensional unfolding is implemented by the program MINIRSA. When used with few data or on data with little variation, however, the solution is not likely to be well constrained, and the algorithm is particularly subject to local minima. It also tends to be expensive in terms of computer time, requiring a large number of iterations to achieve satisfactory improvement. Despite these inherent problems of non-metric unfolding, MINIRSA is generally a useful program so long as the data are sufficient (at least 30 subjects and 8 stimuli is a rule of thumb for a 2-dimensional solution), with variation in the ranks of each stimulus.

Once again, the program can be illustrated by reference to the Bollen-Delbeke data. MINIRSA minimises stress$_2$ and the overall values for 3 dimensions was 0.069, and 0.124 for 2 dimensions. The 2-dimensional configuration is presented in Figure 6.10.

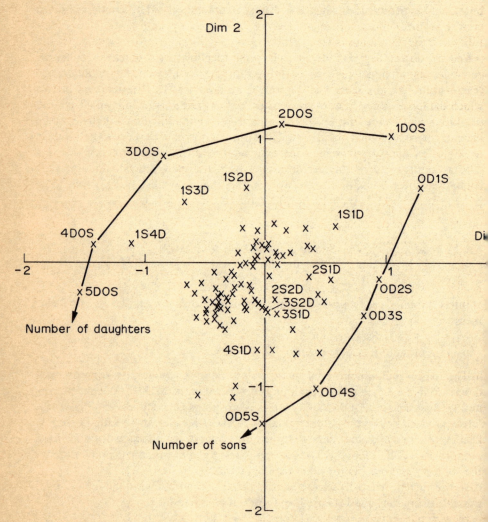

Figure 6.10   *MINIRSA analysis of family size and composition data*

Because it is an internal analysis, it is important to begin by examining the subject points—located in a main cluster at the centre of the configuration (representing preference for largish mixed families), with stragglers to the SE (representing those who prefer a preponderance of boys).

Significant information is additionally conveyed by the occurrence of empty spaces in a configuration. In the vector model, we saw that the fact that parts of the unit circle (or sphere) do not have subject vector ends indicates that no preferences increase in these directions. In the distance case, we need to take note of the *regions* within which ideal points do not occur. This can be done explicitly by constructing the isotonic regions from the stimulus configuration (see Figure 5.5) and then looking at the regions in which ideal points are concentrated and at the regions which do not contain any ideal points.

The stimulus configuration for these data is quite different from the MDPREF one, but can be interpreted in a similar manner. First, the stimuli do form a distorted version of the defining lattice of Figure 6.7a, though the component dimensions (drawn in the figure) are not quite linear or at 90° (uncorrelated); rather, they are pulled in towards one another to enfold the ideal points. Nonetheless, a linear PROFIT fits both 'number of sons' and 'number of daughters' properties with a correlation of around 0.93. The main distortion in the configuration is the way in which the larger family-sizes equally mixed in composition virtually collapse onto one point, and the program deals with them by locating them as close to as many ideal points as possible. Here is a further example of how internal analysis is likely to contort a stimulus structure to satisfy the subjects' data better.

What of the three characteristics of the data? How can these be read out of the configuration?

(i)   *All-boy families are preferred to all-girl families.*   This is evident in the way in which a number of ideal points congregate closely to the Number of Sons line, whilst none are located very close to the Number of Daughters line.

(ii)   *Mixed sex composition is preferred.*   The main concentration within the swarm of subject points lies almost exactly at 45° counter-clockwise inclination, which represents the mixed composition line of the stimuli.

(iii)   *Within a given family size, a preponderance of boys is preferred.*   This also is discernible in the location close to the Number of Sons line.

Although the form of representation is different, the same content can be read out of both the vector and the point representation, but as we have seen, the researcher has to pay especial attention to the *joint* space and the assumptions of the model; the stimulus configuration is not likely to be accurately recovered where there is not enough variation in the data. In any event, it would be advisable to obtain a separate estimate of the stimulus configuration and then map preferences into it, particularly in the case of data based on human judgments of this sort.

6.2.4   *External mapping by the distance model* (PREFMAP-III)

Concisely: PREFMAP (PREFerence MAPping) (Phase III) provides:
   external analysis of two-way, row-conditional data
   by a simple Euclidean distance model,
   using either a monotonic or a linear transformation of the data.

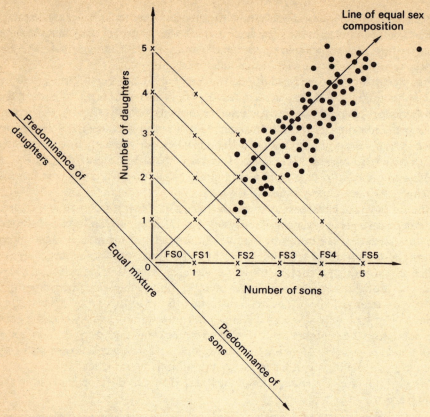

Figure 6.11   *PREFMAP-III (external distance model) analysis of family preference data*

In section 4.4.2 we introduced the idea of representing a ranking (or set of ratings) as a *point* in a configuration in such a way that the relative distances from the new point to the existing ones correspond to the data ranking. Such data might consist of a set of preference rankings or ratings, which are embedded as a set of points (of maximum preference) in an already existing configuration, obtained, say, by scaling an independently-obtained set of data on the similarities between the same objects. This model is implemented in PREFMAP-III and allows for either linear or monotonic transformation of the data. It is also an external analogue of Coombs' multidimensional unfolding model (Coombs 1964, pp. 140–9) discussed in the last section.* As in other instances of external scaling, the fixed input configuration could come from any source: a configuration obtained in previous research (to make findings compatible); a configuration obtained from a *sample* of

*Yet more general distance models, forming PREFMAP-I and PREFMAP-II, are discussed in the next chapter. Since the PREFMAP program implements an hierarchical family of four models of decreasing generality, users may wish to map their data externally according to more than one model. This is effected very simply by choosing a starting phase or model by the S-PHASE parameter and the ending phase by E-PHASE. The program includes approximate ANOVA tests to help the user decide on the appropriate level of complexity. If the user decides to start at a higher level than III, she should be aware that points may have a 'pessimal' or least preferred value on one or more dimension. The meaning of such negative weights is discussed in Carroll (1972, pp. 121–3).

one's subjects (to allow a larger number of subjects' data to be scaled); a 'rational configuration' implicit in the stimuli (as in the Bollen-Delbeke data), and so on. In this case, the current example can be continued, using the configuration of Figure 6.7a as the framework for analysis. Here the prime intervals (difference between N and $N + 1$ sons/daughters) are treated as equal in size and the stimuli are spaced so that both sons and daughters have the same interval width. The resulting mapping of the full 80 subjects is presented in Figure 6.11.

As can readily be seen, the vast majority of subjects prefer families of mixed (and almost equal) sex composition, with differences in preference for overall size of family. This we already know; what the rational configuration helps us see in addition (because of the equal interval spacing) is that there is far greater variation in family *size* preference (the line of equal sex composition) than in the *composition* of the family (the line of mixture/predominance). The fit of the data is also very good. For this metric solution, the overall root mean square correlation between data preferences and distances of the solution is 0.94 (compared to 0.64 for the metric PREFMAP-IV external vector model). There can be little doubt that the simple distance model fits the data well.

Three-Way and Further Extensions of the Basic Model

There are not three incomprehensibles, nor three uncreated: but one uncreated and one incomprehensible

QUICUNQUE VULT (Creed of St. Athanasius)
Book of Common Prayer, 1662

7.1 Introduction

The remaining programs in the MDS(X) series are either designed for the analysis of three- (and in the case of CANDECOMP, higher-) way data or (as in the case of PREFMAP I and II) are more complex variants of models already encountered in Chapter 6.

The main differentiating characteristic of the programs considered here is the form of the model, or rather models, since both PREFMAP and PINDIS consist of a hierarchy of models of increasing complexity. As in the previous chapter, we shall begin by examining the type of data input to these programs—this provides the best clue to their most fruitful areas of application—and then go on to describe the form of the models employed.

7.1.1 Three- (and higher-) way data

The term three-way data refers to a 'cube' of data (see Figure 7.1, p. 192). Such data occur frequently. The third way usually consists of a set of individuals, occasions, methods, points in time, experimental conditions or geographical locations. Two types of 3-way data are usefully distinguished:

(i) **3-way data which are two-mode,** i.e. consist of a set of ordinary 2-way, one-mode, (dis)similarity matrices, and

(ii) **3-way data which are three-mode,** representing for instance the preferences of a set of subjects (mode 1) for a set of food items (mode 2), where the judgments were made at a number of different occasions (mode 3).

Examples of such 3-way data are:

(a) *Three-way, two-mode data* (A set of pairwise (dis)similarity matrices)

A set of individuals (mode 1) each produce a matrix of pairwise similarity ratings between stimuli (mode 2).

Over a number of weeks (mode 1) the mutual attraction of a set of fraternity members (mode 2) is assessed by averaging the preference scores they give to each other.

A set of individuals rate a set of concepts in terms of a set of semantic differential scales. The ratings between each pair of scales are then correlated. This gives rise to a set of correlation matrices between scales (mode 1), one matrix for each concept (mode 2). (Note that in this case what were originally 3-way, 3-mode data have been reduced by the researcher to 3-way, 2-mode data by aggregating over individuals.)

The frequency with which each pair of plant species (mode 1) co-occurs is tabulated for each of a number of locations (mode 2).

A set of attitude items (mode 1) are rated on a 7-point scale by a set of subjects, and a number of different coefficients of ordinal association (mode 2) are calculated between the items.

A set of five live fish (mode 1) are confronted with different stylised shapes of fish, differing on sexual and other characteristics. Their behaviours are summarised in each case by a matrix of rank correlations representing the similarity of behaviour when presented with stylised fish *i* as opposed to stylised fish *j* (mode 2).

(b) *Three-way, three-mode data* (Three distinct sets of entities)

A set of individuals (mode 1) rate a set of automobiles (mode 2) on a set of rating scales (mode 3).

Members of a social group (mode 1) rank each other (mode 2) in terms of emotional closeness. Data are collected on a number of occasions (mode 3).

The input (mode 1)-output (mode 2) matrices between a set of industries is collected for a set of nations (mode 3).

(c) *Higher-way data* (*N*-way data)

There are examples in the literature of four-way data, e.g. semantic differential experiments on a number of occasions (mode 1), using the same set of individuals (mode 2) to judge the same set of concepts (mode 3) on the same set of rating scales (mode 4). (This is 4-way, 4-mode scaling.)

Each year (mode 1), a (different) set of individuals make pairwise judgments of similarity between a set of names of nations. The investigator wished to distinguish European, North American, Latin American and Third World subjects' judgments, and therefore produced a separate correlation matrix between nations (mode 2) for each sphere of origin (mode 3). This is 4-way, 3-mode data.

In principle, data of any way can be scaled, and the CANDECOMP program accepts up to seven-way data. Users are advised to proceed beyond three-way data with considerable caution. They are in largely uncharted territory.

7.1.2 Organisation of the chapter

The defining characteristics of the MDS(X) programs for analysing three-way and related data are described in Table 7.1. As in previous chapters, characteristics of the *data*, scaling *transformation* and *model* are used to define the programs involved. The models described in this chapter consist mainly of generalised versions of the distance and vector models encountered in Chapter 6. The exact form of the generalisation is specified in Table 7.1 under the headings of dimensional weighting and rotation (in the case of distance models) and vector weighting and translation (in the case of vector models). It will be easier to discuss these increasingly complex transformations in the context of the program(s) where they occur.

Let us first take the programs in the order in which they appear in Table 7.1, an order determined by the type of data they analyse. In the subsequent sections

DATA	MODEL	T	R	V	W	P	TRANSFORM**	MDS(X) PROGRAM	Description
TWO-WAY 2-MODE	DISTANCE		R		W	P	M or L	PREFMAP (I) (PM 1)	Dimensional Salience with Idiosyncratic Orientation; Ideal Point.
	DISTANCE				W	P	M or L	PREFMAP (II) (PM 2)	Dimensional Salience with Ideal Point.
THREE-WAY 2-MODE (CONFIGURATIONS)	DISTANCE				W		L	INDSCAL –S	Dimensional Salience
	DISTANCE		(R)				Similarity (S)	PINDIS (PO)	Basic: Procrustes Rotation (General Similarity)
	DISTANCE				W		S	PINDIS, (P1)	Dimensional Salience
	DISTANCE		(R)		W		S	PINDIS (P2)	Dimensional Salience with Idiosyncratic Rotation.
	VECTOR			V			S	PINDIS (P3)	Perspective (Fixed Origin)
	VECTOR	(T		V			S	PINDIS (P4)	Perspective with Idiosyncratic Origin
	MIXED			V	W		S	PINDIS (P5)	Double Weighting
N-WAY N-MODE	VECTOR			V			L	CANDECOMP	N-way Scalar Products

*Model Specification:
Individual: Translation of origin
Rotation of axes
Vector weighting
Weights (dimensional)
Point representation of subjects

**Transform:
Monotonic
Linear
Similarity

Table 7.1 Analysis of 3-way and related data by MDS(X) programs

of the chapter, by contrast, the order will proceed from the simplest to the more complex models.

We have already encountered PREFMAP as a program for mapping two-way, two-mode data into a user-provided configuration according to the vector and simple distance models (see 6.2.1 and 6.2.4). In this chapter these models are extended to include a weighted distance model (PREFMAP-II) and a rotated and weighted distance model (PREFMAP-I). As before, these are chiefly used to analyse sets of preference or, in general, similarity rankings or ratings when the user wishes to represent both the stimuli and subjects in the same solution.

The most common form of three-way data is two-mode, and the most popular form of analysis is the INDSCAL model. This model interprets differences between the subjects (third-way) as arising from differences in the weights (interpreted as importance or salience) ascribed to the dimensions of a common configuration. Because of its conceptual simplicity it makes a natural starting point for discussing more complex models, and is explained in 7.2.1.

An alternative approach to studying individual differences is to scale each matrix separately as an initial stage and then compare the configurations obtained. The PINDIS hierarchy of models provides a successively complex set of models for comparing configurations. There is no reason why the configurations should be obtained in this way; any set of configurations referring to the same set of objects, however obtained and of whatever dimensionality, may legitimately be input. The PINDIS models are discussed last, in section 7.4.1, due to their greater complexity.

This chapter also deals with three-way, three-mode and higher-way data, which may be analysed using a generalisation of the scalar products or factor models already encountered in the last chapter (e.g. MDPREF). The basic ideas of canonical decomposition, used to implement these models, are discussed in 7.2.2, following the exposition of the INDSCAL model which turns out to be a special case.

7.2 Individual Differences and Dimensional Salience

Three-way, two-mode data appear very frequently in the form of a set of (dis)similarity matrices. A typical example occurs when psychologists have subjects make pair-comparison estimates of the similarity between stimuli (such as colour chips) and wish to examine how individuals differ among themselves in the way they perceive colour. (This is the origin of the acronym: INdividual Differences SCALing.) Sociologists often have correlation matrices between a given set of variables for a number of different survey subgroups, and wish to see how the subgroup matrices differ (see 7.2.1.3). Plant ecologists may have co-occurrence matrices for a number of species, one for each of a number of sites chosen to differ on given criteria, and wish to inspect the differences between the sites.

Each example poses a similar methodological problem of aggregation. If the data for each element of the third-way differ to a substantial degree then there is little communality and it is hard to see how they are to be compared at all. If, by contrast, subjects differ in no systematic way but simply represent minor random fluctuations, then there is no point in making anything of the differences. However, if the data are simply pooled together as a single matrix at the outset then all information about differences—whether systematic or random—is lost.

Drawing on ideas developed by Horan (1969), Carroll and Chang (1970) propose the following way of thinking about such individual differences. Suppose each individual (group, or element in the third-way) makes use of a variety of attributes or dimensions in judging the stimuli (the exposition is easiest in psychological terms, but the model generalises easily to encompass any other sort of third-way element). Then define a master or *group space*, which consists of all the dimensions which the subjects happen to use. Each individual subject's space can now be thought of as a special case of the group space—as a reduction of the group space, since she is using some subset of the total available dimensions. This is termed the subject's 'private space'.

In Horan's original formulation, every individual was simply thought of as either using, or not using, each group space dimension, so each 'private space' could be represented by a sequence of 1s and 0s, indicating whether the subject used (1) or did not use (0) the dimensions of the group space. This 'all or nothing' approach was modified by Carroll and Chang by postulating that each subject attaches a *varying* (positive) weight to each dimension which represents the *degree* of salience (or importance, or attention or relevance or centrality) of that dimension to her judgments. So each individual i can be thought of as having an idiosyncratic set of weights, symbolised by $w_a^{(i)}$: the weight given to dimension a of the group space by individual i. These weights hence represent the way in which the subjects differ in the importance attached to each of the dimensions. An individual who attaches equal importance to each of the dimensions will have a set of weights of the same value, and it is such a subject whom the group space actually represents. Others by contrast will attach different weights to different dimensions of the group space and thus systematically distort the group space into the 'subjective metric' of their own private space.

The INDSCAL model presents a way of interrelating these 'private spaces' and provides one-way of comparing how subjects (or elements of the third-way) differ among themselves, but only, be it noted, by accounting for the individual differences in terms of differing weights being associated with the *same* dimensions: INDSCAL is explicitly a *dimensional* model.

7.2.1   *The* INDSCAL *model* *

The Carroll-Chang model is described in full in their definitive 1970 article. A lucid and extended exposition, relating INDSCAL to other forms of three-way scaling is given in Carroll and Wish (1973) and a wide range of applications is discussed in Wish and Carroll (1974). Elementary treatments are given in chapter 4 of Kruskal and Wish (1978), in Spence (1978) and in the MDS(X) documentation. In this section we shall concentrate chiefly on the basic characteristics of the model and upon the interpretation of an INDSCAL-S solution. Further details of the estimation procedure in INDSCAL-S are contained in Appendix A7.2.

Before using INDSCAL-S or embarking upon interpretation of an INDSCAL solution, it is essential to understand clearly the characteristics of the *group space*, the *subject space*, the *private spaces* and their interrelationships.

---

*Hereafter, INDSCAL refers to the model and INDSCAL-S to the version of the program in MDS(X).

(i) The *group space* (denoted **X**) consists of a configuration of p stimulus points in a user-chosen number of dimensions r. The orientation of the axes of this space are *uniquely determined* in the sense that any change in their orientation destroys the optimality of the INDSCAL solution. The INDSCAL axes are often found to be readily interpretable.

This group space acts as the 'reference configuration' to which all the subjects' private spaces may be referred and from which they may be all derived. The group space need not in fact describe any actual subject, and the configuration should not itself be interpreted if it turns out that it is simply a compromise between the configurations of groups of subjects with very different patterns of individual weights.

(ii) The *private space of each subject* i (denoted $\mathbf{Y}^{(i)}$) is a configuration of the p points in r dimensions. Within each private space, distances between stimuli are straightforwardly Euclidean.*

(iii) The *subject space* (denoted **W**) is simply a useful graphical way of comparing subjects in terms of their sets of dimensional weights. It has the same dimensions as the group space and each subject is represented by a vector located by the value of the weights on each of the dimensions.

These basic ideas are illustrated in Figure 7.1 by reference to a simple artificial 2-dimensional example (see also Carroll 1972, p. 105 et seq., Carroll and Wish 1973, p. 57 et seq., and Kruskal and Wish 1978, p. 61 et seq. for similar examples). In this expository example, there are 3 objects and 16 subjects, so the data would consist of 16 lower triangular matrices between the 3 objects. The overall 2-dimensional group space configuration, **X**, consists of 3 points which make an equilateral triangle (representing equal distance between the objects). The private spaces, $\mathbf{Y}^{(i)}$, for subjects 1 and 2 are also presented. Note that in the private spaces the configuration of points no longer forms an equilateral triangle but rather an isosceles triangle (two sides remain the same length but the third is foreshortened). Clearly, *the distances between stimulus points are different within each private space.* The two private spaces are nonetheless related: they may be derived from the reference group space by a simple process of differentially stretching or shrinking the axes of the group space by the square root of the subject's 'importance weights'. In other words, the co-ordinates in the private space (say, for subject 1) are simply a weighted version of the group space co-ordinates. To obtain subject i's private space, we take the co-ordinates of the p stimulus points on the 1st dimension of the group space (x_{ji}) and rescale (stretch or shrink) them by the square root of subject i's weight for this dimension $(\sqrt{w_a^{(i)}})$; that is,

$$y_{ja}^{(i)} = \sqrt{w_a^{(i)}}\, x_{ja}$$

Then the distance between the stimuli j and k in subject i's private space will be:

$$d_{jk}^{(i)} = \sqrt{\sum_a \left(\sqrt{w_a^{(i)}}\, x_{ja} - \sqrt{w_a^{(i)}}\, x_{ka} \right)^2}$$

---

*In INDSCAL the private space of each subject is *estimated* as a distortion of the group space directly from the data. In PINDIS, by contrast, each subject's 2-way data are *first scaled* and then input in the form of configuration co-ordinates into the program.

or, in simplified form (taking the weight outside the squared term):

$$d_{jk}^{(i)} = \sqrt{\sum_a w_a^{(i)} \left(x_{ja} - x_{ka} \right)^2}$$

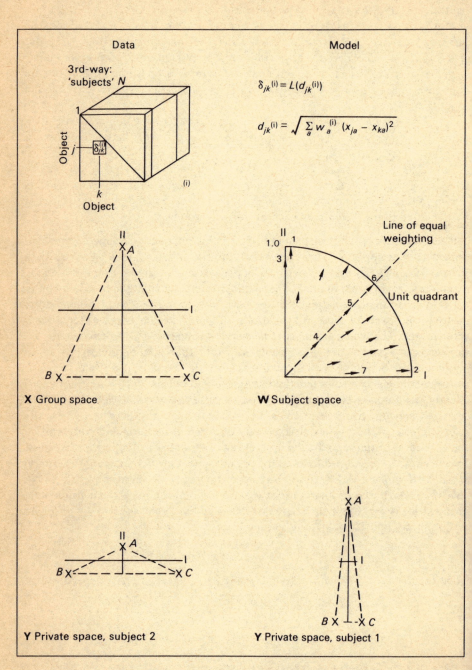

Figure 7.1   *Basic INDSCAL model*

This last equation gives the general form of the INDSCAL *and other weighted distance models:* A 'subject' i's judgment of the (dis)similarity between objects j and k is taken to be a (linear) function of the overall distance between stimuli j and k in the group space, after that space has been differentially rescaled (stretched and shrunk) by the subject's set of weights into the 'subjective metric' of the subject concerned.

7.2.1.1   *The group stimulus space: its properties and interpretation*

The group stimulus space functions as the basic reference configuration from which the private configurations of individual subjects can be derived by differentially shrinking or stretching the dimensions by the (square root) of the corresponding weights.

The INDSCAL dimensions actually represent the (orthogonal) directions where the variation among subjects is the greatest: it is for this reason that they are normally easy to interpret. These dimensions are uniquely identified, in the sense that if the original dimensions are rotated and new subject dimension weights calculated, the resulting solution will explain the subjects' data less well than the original solution.† If it turns out that the extent of individual differences is not great, then such a reduction in explained variance is likely to be small, but in the normal way the reduction is usually fairly substantial. Unless there are compelling reasons of interpretability or little subject variation, the INDSCAL axes should be regarded as fixed.

In most MDS solutions encountered so far, the final configuration is rotated to principal axes—that is, dimensions are chosen which have the statistically convenient property that co-ordinates of the points are not correlated across dimensions. *This is not (generally) true of an* INDSCAL *group space:* the dimensions of greatest subject variation will usually give rise to a configuration where the co-ordinates of the stimulus points are to a greater or lesser extent correlated. ‡ The information about the extent of this correlation between pairs of axes is contained in the output from INDSCAL-S in the matrix of scalar products between dimensions ('sums of products') for matrix 2.

The INDSCAL group stimulus space configuration should therefore be interpreted with caution: strictly speaking it represents a subject who weights the dimensions equally, and if a significant number of subjects' weights depart markedly from equality then there is a danger of trying to interpret a configuration which is in no sense representative. That said, methods for external interpretation of INDSCAL *dimensions*—and especially linear property-fitting (see 4.4.1)—are particularly appropriate, since the dimensions are *not* arbitrary and it is important to try to tie down their meaning as accurately as possible. Good examples of the use of

†The unique orientation of axes in the INDSCAL model means that the solution is unique up to permutation of axes, which is equivalent to saying that the only permissible rotation of the dimensions which preserves all significant information is through multiples of 90°. However, the actual size of the configuration is arbitrary, and is therefore normalised so that the variance of the projections on each of the co-ordinate axes is unity and the centroid of the configuration provides the origin (Carroll and Wish 1973, p. 30).

‡An option SOLUTIONS (1) exists in the MDS(X) version to obtain a solution where the axes are as close as possible to being uncorrelated. Such a solution will normally be sup-optimal compared to the ordinary solution.

property fitting to validate or confirm the interpretation of INDSCAL dimensions occur in the classic Carroll and Chang (1970) paper and elsewhere.

INDSCAL-S can also be used in an external mode if the user provides the program with a group stimulus space configuration (which remains fixed in orientation) and the INDSCAL analysis then concentrates entirely upon estimating from the subjects' data the subject weights for this configuration. (External use is achieved in INDSCAL-S using the FIX POINTS (1) option). External analysis of this sort has two main uses: (i) to scale a large number of subjects' data and (ii) to compare a number of different data sets by referring them to a common reference configuration. Thus if the user has, say, 500 matrices for analysis, it is sensible to choose a manageable sample of those matrices and scale them. The resulting group stimulus space can then be fixed, and the subject weights can then be estimated for as many batches of subject matrices as desired.* An example of the second use occurs where a replication has been made of a previous study and the researcher wishes to investigate the extent to which her subjects' data compare to the weights obtained in the earlier study. The original group space configuration is fixed under this option, and the subjects' weights may then be estimated and compared to those of the original study.

7.2.1.2   *The subject space: its properties and interpretation*
When subjects' data are input to INDSCAL-S they are normalised to have equal weight, which has the effect of giving each subject's data equal influence on the solution. This fact, in conjunction with the normalisation of the group stimulus space described above, gives rise to several nice properties of the subject space which are useful to bear in mind when interpreting an INDSCAL solution:

   (i)   The subject's weight on a dimension is (approximately) equal to the correlation of the intervals between stimulus co-ordinates on that dimension and the corresponding pairwise dissimilarity values in the subject's data
   (ii)   Consequently, the *squared* subject's weight on a dimension is (approximately) equal to the proportion of variance in the subject's data that can be accounted for by that dimension (Wish and Carroll 1974, p. 452).
   (iii)   Therefore, the squared distance from the origin of the subject space to a subject's point in that space is (approximately) equal to the proportion of variance in the subject's data accounted for by the full INDSCAL solution.

If the dimensions of the INDSCAL solution are uncorrelated, then the word 'exactly' replaces the word 'approximately' in the above three sections. Thus in the subject space portrayed in Figure 7.1, subjects 4, 5 and 6 provide an example of subjects who weight the dimensions equally; they differ only in the fraction of their data explained by the model, with the data of subject 6 perfectly accounted for. Similarly, subjects 7 and 2 have the same pattern, giving virtually exclusive salience to dimension I, whilst subject 3 uses only dimension II. Looking at the pattern in terms of goodness of fit, the data of subjects 1, 6 and 2 are totally accounted for, whilst those of subject 4 are very poorly explained.

*See Coxon and Jones 1979, pp. 54–9, and especially T3.17, for an example using a balanced set of 68 matrices to obtain the group space configuration by reference to which 286 subjects' subject weights were estimated.

Note that only *positive* weights are allowed by the INDSCAL model. If, as occasionally happens, a very small negative weight occurs in may be considered as approximation to a zero weight; if it is substantial it can only be interpreted as indicating that the basic model does not hold for the data of the subject concerned.

The significant information in the subject space is contained, then, (1) in the *direction* in which a point is located from the origin, since any points lying on line from the origin have weights in the same ratio, and (2) in the *distance* (of a subject vector) from the origin, representing how well the subject's data are explained by the model.

Before embarking on any systematic analysis of INDSCAL subject weights, it is important to know something of the stability of INDSCAL solutions (see Jones and Waddington 1973; MacCallum 1977).

(i)   Simulation studies show that, even in circumstances of high error in the data, recovery of the group space configuration and its dimensional orientation is excellent, but that

(ii)   the stability of the subject space is far less stable and much more subject to fluctuation in the presence of error.

The temptation to use cluster analysis on subject weights should be strongly resisted: the separations of subject points are in no sense ordinary distances and their location is far from stable. The question of whether any linear procedures such as ANOVA and its multivariate variants should be used on INDSCAL weights remains contentious. MacCallum (1977) and others often strongly counsel against their use; Carroll and others think that a more lenient approach is called for.

Usually the user will want to compare subjects in terms of the patterns of relative salience given to dimensions. This is best done by concentrating on the angular separation between subject vectors: the smaller the angle of separation, the more similar is the pattern of weights. In the two-dimensional case, it is usually a simple matter to see closely collinear 'sheaves' of subject vectors in the subject space, and such bunching can also be detected visually in three dimensions. Beyond that, statistical analyses of different subject vectors should be used (see Mardia 1972; Coxon and Jones 1979, pp. 128–36 for use in an MDS context). A simple alternative for two-dimensional data is simply to take the ratio of the weights for each subject and, since the distribution of such ratios is usually markedly positively skew, it often makes sense to correct this by taking a logarithmic transformation of the weight ratios.

An alternative to Carroll and Chang's representation of subject weights has been suggested recently by Young (1978). The Young Plot allows the amount of variation explained to be represented independently of the relative salience of the subject weights, and is illustrated in Figure 7.3b (p. 199).

The Young Plot
The Young Plot charts each subject in terms of two things—the relative salience ascribed to one dimension over another (on the horizontal axis) and how well the subject's data are fit by the model (the vertical axis). The first is measured by the ratio of the two-dimensional weights—which can be interpreted trigonometrically

as the tangent of the angular separation between a subject's vector and the line of equal weighting in the conventional representation of INDSCAL subject space.* The goodness of fit is given simply by the squared correlation r^2 between the subject's original data and the values predicted by the INDSCAL model. This information is provided separately in an INDSCAL run.

The Young Plot and its construction from a set of subject weights is illustrated in Figure 7.3b and is very simple to read. Subjects located in the centre of the horizontal axis (such as the Labour group of voters in this example) weight the dimensions equally; the more that dimension I dominates over II the further left the subject point is, so the non-voters group has the most dominant weight for dimension I and the Conservative voters group has the most dominant weight for dimension II. The goodness of fit is simply read up the vertical axis. In this example the greatest differentiation is between the 'other parties' group whose data are not well explained (being largely Scottish and Welsh Nationalist party supporters they are presumably dancing to a different piper) and the others.

The most important advantages of the Young Plot are that it gives accurate representation of patterns of dimensional weighting and of goodness of fit independently of dimensional correlation, and concentrates the user's attention onto the angular separation (relative salience) of patterns of subject weights rather than on the proximity of points portrayed somewhat misleadingly in a conventional subject space. The Young Plot can also be modified in various ways—to portray patterns of three-dimensional weights, or to compare relative salience of weights with any other variable of interest (see Coxon and Jones 1980, p. 59 et seq.).

7.2.1.3 An example: political party imagery

Alt et al. (1976) carried out a survey of 2,462 British voters after the 1974 British election. The questionnaire included 20 attitudinal items—political party features (items 1–7), the parties' handling of contentious issues (items 8–10), blame (11–12), taxes and pensions (13–14) and policy positions (15–17). These are reproduced in Table 7.2. Each pair of items was cross-tabulated and the association between them measured by Goodman and Kruskal's gamma, which preserves weak monotonicity of the item categories (see 2.2.2 above). The respondents were divided into five subgroups, viz

A   Conservative voters
B   Labour voters
C   Liberal voters
D   Other voters (principally Scottish and Welsh National Parties)
E   Non-voters

Each of these subgroups were then treated as a 'pseudo-subject', and gamma coefficients were calculated for each subgroup, hence providing a (5 × 20 × 20) array for input to INDSCAL. The group space configuration is given in Figure 7.2 and

---

*The tangent of the angle which the subject vector makes with the first dimension (tan θ_1) is defined as the ratio of the weight on dimension II to the weight on dimension I. Tan (θ_1 − 45°) measures this predominance of dimension II over dimension I as a deflection (angular departure) from the line of equal weighting.

Item No.	Symbol	Title
1	K	Keeps/breaks promises
2	D	Divides/unites country
3	B	Bloody-minded/reasonable
4	G	Good for one/all classes
5	E	Extreme/moderate
6	Ca	Capable/not capable
7	SF	Stands firm/gives way
8	P	Prices
9	M	Miners' strike
10	S	Strikes
11	PB	Blame for prices
12	MB	Blame for miners' strike
13	T	Taxation
1	Pe	Pensions
15	CM	Common Market
16	N	Nationalisation
17	SS	Social services
18	W	Wage controls
19	C	Communists
20	R	Reliability

Table 7.2   *Items in political party imagery study* (Alt et al. 1976) (Reproduced by permission of the journal *Quality and Quantity*)

the subject weight plots are given in Figure 7.3. Alt et al. identify dimension I as 'image consciousness' (by which they mean an ideologically-based concern with both political style and performance) and dimension II as 'policy consciousness' (concerned primarily with welfare and related policy issues). Note from the shape of the group stimulus space configuration that the two dimensions are clearly positively correlated. The authors do not provide this information, but our estimate is $r_{I, II} = 0.23$.

Further interpretation of the group space should wait upon inspection of the subject weights (Figure 7.3). Even a cursory examination of the subject space (a) and more obviously of the Young diagram (b), shows very considerable differences in the goodness of fit and in the relative salience of two-dimensional weights between the subgroups.

But just how significant are these relative differences in weights, given what we know of the relative instability of INDSCAL weights? Alt et al. use an unusual form of internal validation. They divide their subjects into a number of pseudo-groups based upon 'irrelevant' factors (such as male/female) and random criteria (exclusive but randomly constructed subgroups and overlapping random subsamples of subjects) and proceed to estimate weights for each group, keeping the reference configuration fixed. Only if the voter subgroup differences exceed the

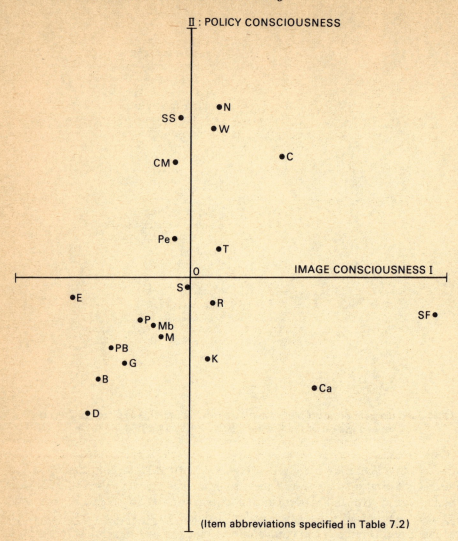

Figure 7.2   *INDSCAL group space: political imagery study*

random differences do they consider them to be sufficient to merit separate treatment. It turns out that the differences among voter subgroups greatly exceed those found for the random groups, especially on the first dimension. The authors then construct the private space for each subgroup (pp. 308–9), and comment:

> The relative unidimensionality of the items for Liberals and Non-voters is apparent. For them, big differences between items only occur between those most clearly reflecting 'style' and 'performance'. In contrast, voters for the two major parties use both dimensions in differentiating items, and the previously mentioned differences between these groups are also evident. Particularly striking is how small the group space looks—how undifferentiated all the items appear—to 'voters for other parties'. These results are substantively not necessarily surprising: the items were, after all, re-scaled as inter-party

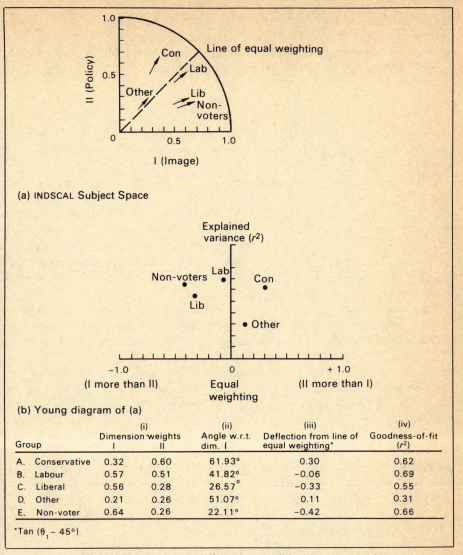

(a) INDSCAL Subject Space

(b) Young diagram of (a)

Group	(i) Dimension weights I	II	(ii) Angle w.r.t. dim. I	(iii) Deflection from line of equal weighting[*]	(iv) Goodness-of-fit (r^2)
A. Conservative	0.32	0.60	61.93°	0.30	0.62
B. Labour	0.57	0.51	41.82°	−0.06	0.69
C. Liberal	0.56	0.28	26.57°	−0.33	0.55
D. Other	0.21	0.26	51.07°	0.11	0.31
E. Non-voter	0.64	0.26	22.11°	−0.42	0.66

[*]$\text{Tan}\,(\theta_1 - 45°)$

Figure 7.3   *Subject weights plots: political imagery study*

comparisons. The clarity and parsimony with which INDSCAL recovers this property of the group space is, nevertheless, impressive.

(Alt et al. 1976, p. 310)

This example of the use of INDSCAL shows well how, with a little initiative and imagination a model developed within an individually-based psychological tradition can be adapted with considerable success to analyse survey data referring to several thousand respondents (see also Coxon and Jones 1977).

7.2.2   *Canonical decomposition* (CANDECOMP)

Canonical decomposition (CANDECOMP hereafter) is a very general model, of which INDSCAL, two-way scalar products (factor, vector) models and multiplicative conjoint analysis are the better-known special cases. The N-way CANDECOMP model states that an N-way, N-mode array of data ($z_{ijk\ldots n}$) can be decomposed into a separate set of numerical 'effects' for each way, which combine multiplicatively within a dimension, and additively across dimensions:

$$z_{ijk\ldots n} \simeq \sum_a v_{ia} w_{ja} x_{ka} \cdots y_{na}$$

where the v_{ia}, w_{ja}, x_{ka} and y_{na} are the numerical 'effects' or 'weight' parameters to be estimated (one weight per dimension), and $\simeq$ means a least-squares estimation or approximation.*

An example of a 4-way CANDECOMP application would be an investigation of the connotative meaning (Osgood 1965) of a set of political concepts, where the investigator had 60 subjects evaluate 12 concepts on 25 bipolar rating scales on 4 different occasions. In this case, having decided upon an appropriate dimensionality for the analysis, the researcher could use CANDECOMP to estimate the numerical effect or weight to be ascribed to each element in each of these four ways, assuming that the weights combined together multiplicatively (in a manner akin to scalar products in the 2- and 3-way case).

A quite common variant of CANDECOMP occurs when two ways of the data (conventionally known as the second and third ways) refer to the same set of stimuli, and the analysis then becomes a higher-way generalisation of INDSCAL. Continuing the earlier example, the researcher might decide to ignore individual differences and run correlations between the 25 rating scales of each of the 12 concepts for each of the 4 occasions, thus forming a 12×4 stack of 2-way symmetric correlation coefficients between the rating scales. Labelling the occasions as the 1st way and the concepts as the 4th way, we now have a ($4 \times 25 \times 25 \times 12$) data array. Since this is an extension of INDSCAL the researcher would normally require that the effects of the stimuli (rating scales) be constrained to be the same, as in other INDSCAL analyses. (This is effected by stipulating SET (1) on the PARAMETERS card, but unless this equalisation condition is imposed, the weights for each separate way will be estimated.)

7.2.2.1   *Normalisation of the data and solutions*

In INDSCAL, each subject's data are normalised before the analysis, so that each subject is given the same influence in the solution. This is not the case in CANDECOMP: any modification or normalisation of the data the user wishes to make must usually be performed before input. In many cases the actual unit of measurement of the data is in any case arbitrary, and may vary from way to way in the case of higher-way data. But since CANDECOMP does not equalise data at the outset, these arbitrary aspects will affect the analysis. Modification of data before input in order to control for arbitrary characteristics can be illustrated by a now famous example drawn from two-way data scaling. Coombs (1964, pp. 464–6) analyses the co-citation patterns of psychological journals: how often articles in Journal X cite articles from Journal Y. Now journals differ not only in the extent to which they

---

*Three-way CANDECOMP is closely allied to Tucker's Three-mode Factor Analysis developed in the mid-1960s and related by him to three-way scaling in his 1972 paper (Tucker 1964, 1972). The interrelationships are also discussed in Carroll and Wish 1973.

make citations, but also in the number of articles they publish—and both of these characteristics may be viewed as irrelevant, arbitrary, aspects of the data if one is interested in studying the pattern of *relative* citation patterns between the journals. He therefore decided to remove both the row effects (representing the differing size of the journals and the conventions about the appropriate number of citations they make) and the column effects (representing the overall popularity or frequency of citations of each journal) by the device of 'double-centring' the matrix. This leaves only the interaction effects to be scaled. (Row and column centring of the input matrix are options within the MDPREF program.) In higher-way data matrices, the same problem arises and a facility is provided for triple- (or higher-) centring the data matrix, using CENTRE (1). This option should normally be chosen, unless the user has reason to keep the original values of the data. (See Gower 1976 and Tagg 1979 for a discussion of this question in the 3-way case.)

The *solution configuration*, by contrast, is normalised in both CANDECOMP and INDSCAL. In CANDECOMP, the solution weights *for all but the first way* are normalised so that the sum of the squares of the weights on each dimension is unity, (Carroll and Wish 1973, pp. 30–1).* Moreover, in CANDECOMP the origin of the spaces is not constrained to be at the centroid, as occurs in the INDSCAL group space, unless the data have been initially centred.

7.2.2.2 *Using* CANDECOMP

Three-way, three-mode CANDECOMP and four-way three-mode INDSCAL are the most obviously useful variants of the *N*-way model. Beyond these the user is advised to proceed with caution, since very little is known about the properties and stability of the solution, although one example of four-way three-mode INDSCAL has been alluded to in the literature (Carroll and Wish 1973, p. 98), where subjects (1) made pairwise similarity judgments between nations (2 and 3) by a variety of methods (4). The authors themselves counsel users to remain with three-way analysis:

> It is not clear when or whether these higher (than three) way models are appropriate. In their current form they impose strong, and usually unrealistic constraints on the data ... In most cases it is more appropriate to simply concatenate the *N*-2 nonstimulus modes into a single mode, and then do a three-way analysis. (For example, if there are n_1 subjects and n_2 conditions there would be $n_1 \times n_2$ 'pseudosubjects' defining the third way of the anlysis; there would be one two-way matrix for each of those 'pseudosubjects'. Another alternative would be to do separate three-way analyses for each of the n_2 conditions).
>
> (Carroll and Wish 1973, p. 99)

The effect of this warning has perhaps been too discouraging: few published examples of CANDECOMP as yet exist.

One of the earliest is presented in Green and Rao (1972, pp. 45–8), where they use 3-way CANDECOMP to assess the congruence between a set of 9 configurations derived from the same data, but each obtained by a differing scaling method. They input the 2-dimensional co-ordinates of 9 scaling solutions referring to 15 stimuli

---

*For this reason (because all differences in the sum of squares are absorbed in the first way) users are advised when using the program to identify the first way with the one which represents for them the most important source of variation.

(breakfast foods), choosing the 9 solutions to be the first way, the 15 stimuli to be the second way and the 2-dimensional co-ordinates of the original solution to be the third way. From an analysis of the 2-dimensional CANDECOMP solution weights for each way, they conclude:

(1)   The nine scaling algorithms represent essentially the same configuration after allowing for differential stretching along the axes.

(2)   The stimulus configuration is a composite of the original configurations, but one where the axes are interpretable and the pattern of points is very similar to the position in the original configuration.

(3)   The weights for the original orientations of the configurations again confirms 'the closeness of the CANDECOMP orientation to the original orientations'.

In a later section we shall see that PINDIS (a program which had not been developed at that time) provides a more precise way of investigating how transformations such as axis-stretching allow the relationship between configurations to be examined more precisely.

Canonical decomposition is clearly a very general and wide-ranging procedure which can also be adapted to estimate the parameters of a wide variety of models, including latent structure analysis.* But its very flexibility and power mean that the user should exercise caution in its use and not let enthusiasm outrun understanding.

7.3   Comparing Configurations

When using MDS, it is only a matter of time before the user wishes to compare two or more configurations. If a study has set out to replicate a previous one, then it is important to know the ways, and the extent, to which the current MDS solution resembles that of the original study. More often, the user has employed more than one variant of MDS on the same data, or has scaled the data of different subgroups of subjects, and the issue of similarity between the resulting configurations once again arises.

Before attempting to compare configurations the user should be sufficiently persuaded that in each case the fit of the solution to the data is good enough to warrant proceeding any further.

The answer to the question, 'How similar are two configurations?' depends on two things:

(i)   what aspects of the configuration are considered *relevant*, and
(ii)   what properties of the configuration are *unique*.

---

*Latent structure is a family of models developed to analyse sets of dichotomous response patterns (Lazarsfeld and Henry 1968; Fielding 1977). In these models, observed patterns are thought of as arising from the multiplicative combination of the latent probabilities of giving a positive answer to each item, where these probabilities differ according to the position of the subject in the space. In the latent class model, the 'space' consists of a set of partitions or classes. Whilst the ideas of latent class analysis are very appealing, estimation of the model parameters have proved to be very difficult. However, Carroll (1975) shows how the latent class model may be seen as a special case of canonical decomposition, where the dimensionality may be identified with the number of latent classes and the parameters of the model may be estimated in a straightforward manner from the CANDECOMP weights. Paradoxically, the CANDECOMP estimation is better than the procedure suggested by Lazarsfeld and Henry, despite the fact that it is minimising a different and less obviously appropriate badness-of-fit function.

For instance, if one configuration were simply twice the size of the other but in every other way identical, it is unlikely that most researchers would consider this a relevant difference. Hence one would normally wish to compare *relative* rather than absolute distances. If dealing with solutions obtained from a Euclidean distance model, it is unlikely that the orientation of the configuration will be considered relevant since any rigid rotation leaves distances unchanged. Similarly, the origin of the space would normally be treated as arbitrary, unless the model were a vector model (since change of origin alters scalar products and hence alters the angles separating the vectors, cf. Appendix A2.1) or unless a facet analysis had been employed in interpretation and the point chosen as the origin was substantively meaningful.

In actual fact, many users of MDS and factor analysis resort to crude and misleading methods for comparing configurations—such as restricting attention to the first two dimensions of a solution and simply 'eye-balling' the configurations, hoping that salient differences will in some way reveal themselves. Unfortunately, differences which are irrelevant can often make identical configurations *appear* to be very different. This is illustrated in Figure 7.4, where the two-dimensional configuration of seriousness of offences given in Figure 1.1a, is first reproduced as Figure 7.4a, and then submitted to a series of transformations which preserve relative distances. In terms of keeping the same relative distances these two configurations are identical but they certainly *look* different. So appearances are not a reliable guide as to how two configurations are alike. How, then, does one go about comparing them?

Sometimes transparent acetate sheets are used to compare two-dimensional configurations—one configuration is first copied onto a transparent sheet and laid

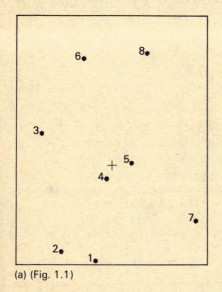

(a) (Fig. 1.1)

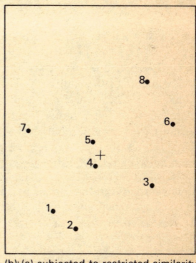

(b): (a) subjected to restricted similarity
transform:
(i) clockwise rotation through 40°
(ii) reflection in axis I
(iii) rescaling by factor of 0.75

Figure 7.4 *Two 'identical' configurations*

on top of the other configuration. By moving the sheet in a circular manner (rotations) and/or flipping it over (reflections) the one configuration can be moved into maximum apparent congruence with the other. This procedure has its uses, but even this expedient cannot deal with differences of scale, and is obviously restricted to two-dimensional situations. Moreover, we still need some explicitly defined index of configurational similarity if the comparison is to be anything more than approximate or if we need to compare more than two configurations. How, then, should two or more configurations be compared?

(i)   First, we must return to the question of what aspects count as significant and unchangeable information and what are irrelevant.

(ii)   That decided, a method is needed which will bring the configurations into the closest possible conformity with each other.

(iii)   Finally, some indication is needed of how closely the configurations correspond to each other, preferably by using an appropriate measure of goodness of fit.

We shall take up each of these points in turn.

First of all, in this section (and indeed, as far as 7.4.1) we will assume that two configurations are considered identical if they only involve differences of:

(1)   *scale* (how large the actual configuration is);
(2)   *orientation* (rigid rotation and/or reflection of axes); and
(3)   *origin* (the zero point of the space).

Consequently, configurations may be shrunk or expanded at will (1), moved—rigidly rotated—through any angle (2), and may have the origin translated to any point in the space (3) in order to get them into greater conformity with each other. The value of any index of similarity between configurations should remain unchanged whenever these operations are performed.

7.3.1   Geometric transformations of a configuration

The geometry of these three basic operations, which taken together define an 'extended similarity transformation', is described in Appendix A7.1. Users who are unfamiliar with them should read it carefully before proceeding further. Two transformations are particularly important in examining the similarity between configurations and in moving them into closest conformity. Both keep the comparative or relative distances in the configurations unchanged:

Configuration transformations which preserve (Euclidean) distances
1   *Extended similarity:* involves rotation, translation and rescaling
2   *Restricted similarity:* involves rotation and rescaling (no translation of origin)

Usually it will not be possible to transform two (or more) configurations into an identical structure, and it is therefore necessary to define an index of how similar two configurations are. All commonly encountered indices can be thought of as being a function of the *distances* between the points in the two configurations. Gower (1979, 1980) has provided an excellent review of such indices by examining the form of the function relating the distances in the two configurations. One obvious measure for assessing the similarity between two configurations is the product moment correlation* between the distances involved, and this is commonly used. A related measure of similarity between the configurations is called S by Lingoes (Lingoes and Schönemann 1974, p. 436). This has the nice properties that its value depends neither upon the number of points, nor upon the number of dimensions, nor on the scale of the configuration. Consequently it can be used to answer the question, 'Does configuration X match Y better than X matches Z?', which will be a central issue when we come to compare several configurations. An even simpler measure, which is just the squared linear correlation r^2 between the co-ordinates of the configuration X and Y which have been brought into maximum conformity, is also frequently used. It is directly related to S and shares its desirable properties.†

7.3.2 Comparison using Procrustes analysis

Procoptes, better known as Procrustes, 'The Smasher', caught (travellers) on the borders of Athens and by stretching or pruning made them an exact fit to his lethal bed.

(Kirk 1974, p. 153)

As originally employed in comparing configurations, simple Procrustes analysis consisted of moving two configuration matrices **X** and **Y** into closest conformity, allowing only rotation and reflection. Procrustes analysis was later extended to include rescaling and translation of origin. This more extended usage, which will be employed here, is termed generalised Procrustes analysis (Gower 1975) and is illustrated in Figure 7.5. Here three configurations (A, B, C) of four points each forming a roughly rectangular shape have been positioned into closest conformity with each other and, since the configurations are not identical, a new configuration **Z** is then produced, defined by the location of the square points in Figure 7.5. This new configuration is an 'average configuration' in the sense that each of its points is a least-squares fit to the corresponding points of the original configurations. This best fitting configuration is called by a variety of names: the compromise, consensus, centroid (as in PINDIS) or group configuration (as in INDSCAL). From now on it is referred to as the centroid configuration, and denoted by **Z**.

The iterative computing procedure for producing the centroid configuration is described in Gower (1975, p. 43).

Procrustes rotation can be used on any number of configurations, and is commonly thus used.

7.4 Procrustes Rotation and Individual Differences Scaling

Procrustes rotation is not only useful for comparing configurations, it can also serve as a basis for individual differences scaling. The basic idea is simple. Given a set of configurations, we begin by moving them into closest conformity by

---

*Carroll (1972) has suggested the use of linear, monotone, non-linear (continuity) and canonical correlation coefficients for assessing different aspects of goodness of fit between two configurations, according to different criteria for assessing 'fit'.

†$S^2 = 1 - r^2(X, Y)$ is the equation relating the two measures, and both measures remain unchanged when a similarity transformation is performed on X and Y to bring them into maximum conformity.

generalised Procrustes analysis, thereby producing the centroid configuration. Next we measure how closely each configuration fits the centroid configuration. Conceptually, the *badness* of fit of each configuration to the centroid can be defined in terms of the sum of squares of the residuals, i.e. $\Sigma (x_i - z_i)^2$ in Figure 7.5, but we shall follow Lingoes' example and use the squared correlation between the 'subject' configuration co-ordinates and those of the centroid configurations as a measure of *goodness* of fit.

7.4.1   *The* PINDIS *models*

Lingoes has developed a series of increasingly complex models based on Procrustes rotation, which have affinities with other models we have encountered. His set of models is known collectively as PINDIS (Procrustean INdividual DIfferences Scaling).

The data for PINDIS, unlike other three-way models, consist of a *set of configurations* obtained from previous scaling solutions. The configurations are first moved into maximum conformity by generalised Procrustes rotation. Since this procedure consists entirely of 'admissible' transformations (in the sense that they leave the relative distances unchanged) this basic general similarity 'model' (denoted P0) provides a yardstick or reference point for subsequent models which do *not* leave original relative distances intact.

Beyond P0, the models all involve so-called 'inadmissible' transformations—

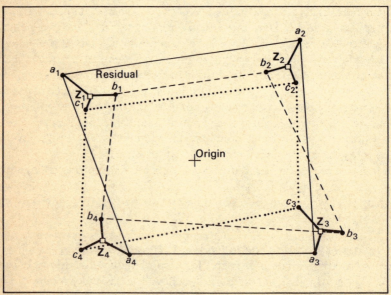

(Adapted, with permission, from Gower 1975)

3 configurations of 4 points, with a common origin: A denoted by ———
                                                          B denoted by – – – – –
                                                          C denoted by ·············
Best fitting Procrustes ('centroid') configuration is given by Z (z_1, z_2, z_3, z_4)
It minimises sum of squares of residuals (denoted by thick lines).

Figure 7.5   *Procrustes analysis of 3 configurations*

i.e. operations which change the original relative distances in some systematic way in order to obtain a better fit between the original configuration $\mathbf{X}_i$ and the reference centroid configuration $\mathbf{Z}$. At this point it should be stressed that:

(i)   the centroid configuration is defined somewhat differently in each model unless the user wishes to keep it unchanged throughout (in this latter case, PINDIS becomes an 'external' analysis, an option effected in the MDS(X) version by using the READ HYPOTHESIS command);

(ii)   at each stage (in each model) the individual configurations are moved into optimal fit to the centroid configuration, using admissible transformations.

The interrelations between the PINDIS models are illustrated in Figure 7.6. A brief simplified resumé will be given here, and separate models are discussed in greater detail in subsequent sections. In the diagram, each model is represented by a box. The top half indicates the usual name of the model, and the bottom half indicates what operations are performed on the centroid matrix ($\mathbf{Z}$) and what 'individual differences' parameters are estimated in the model. Arrows are drawn upward from less general to the more general models. Note that the models do *not* form a strict order, but rather two parallel hierarchies, i.e. the *distance* models (P2, P1, P0) and

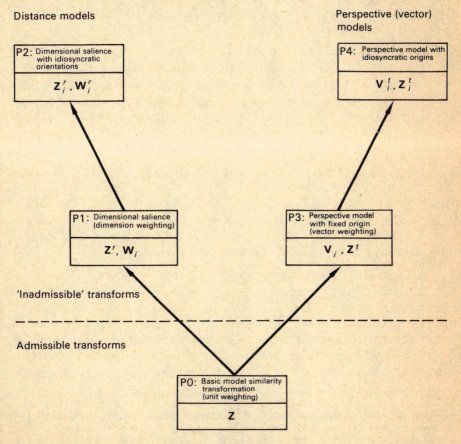

Figure 7.6   *PINDIS models*

the *perspective/vector* models (P4, P3, P0), which share a common basis (P0).*

Before going on to describe the models, some preliminary points need to be made about the form of their specification (in the lower half of the boxes in Figure 7.6).

(i)   The parameters of each model are of two sorts—they refer either to what is done to the centroid matrix, $\mathbf{Z}$, and to what systematic weights (dimensional weights, $\mathbf{W}$ or vector weights, $\mathbf{V}$) are applied to move the centroid configuration into greater conformity with the original configurations.

(ii)   The irritating superscripts and subscripts which bedeck the matrices in the model specification are in fact necessary to distinguish between the models, and repay careful attention. The original average centroid configuration $\mathbf{Z}$ appears in its pristine, undecorated form in P0. In every other model the centroid configuration is changed in some way.

In the distance models, the superscript r signifies that it is rotated. Thus the centroid configuration is rotated in both P1 and P2—to a different orientation for each individual configuration in P2 (signified by $\mathbf{Z}_i^r$) and to a *single* new orientation in P1 (signified by $\mathbf{Z}^r$), indicated by the absence of a subscript. In the vector models, the superscript t denotes translation or shift of origin. Hence the centroid configuration is translated to a unique position for each individual in P4 ($\mathbf{Z}_i^t$) and to a single new origin in P3 ($\mathbf{Z}^t$).

The Distance Models

P1   The dimensional salience model is the PINDIS equivalent of INDSCAL. The centroid configuration is first rotated into an optimal position for dimensional weighting and then a set of individual dimension weights are estimated for each input configuration. These weights are entirely analogous to INDSCAL weights, except that in PINDIS they may legitimately be negative, in which case they signify the reflection of the dimension concerned.

P2   The individually rotated dimensional salience model is the PINDIS equivalent of the Carroll-Chang (Carroll and Wish, 1973, pp. 90 et seq.) IDIOSCAL model which allows the axes of the centroid configuration to be rotated to an idiosyncratic orientation for each individual configuration, and then be differentially weighted. In this model, the dimension weights will only be comparable between 'subjects'/configurations if they happen to share the same rotation.

The Vector Models

P3   In the basic simple perspective model, the origin of the centroid configuration is first translated to an optimal position and a vector is then constructed from the origin to each of the constituent stimulus points. Each individual configuration can be thought of as having had a different set of weights applied to each of the vectors. A high vector weight will push a point further out from the origin, and a low vector weight will contract the vector and move a point closer to the origin. A negative

---

*A further 'double-weighted' model is discussed by Lingoes and is available in the PINDIS program. However, this model is particularly subject to sub-optimal solutions and its use is not generally recommended. Its estimation is suppressed in the MDS(X) version by setting SUPPRESS(1). The double-weighted model is not treated further here.

weight (which is quite permissible) has the effect of 'flipping' the vector in the opposite direction. The effect of a set of vector weights is thus to 'unscramble' a configuration by selectively relocating the stimulus points of the centroid configuration, with the proviso that they can only move in the same direction, towards or away from the origin.

P4 The individually translated perspective model differs from P3 in allowing *each* individual configuration to have its own 'point of view' (origin). Since, as we know, translation of origin changes vector separations, the same set of vector weights may well have markedly different effects on differently centred configurations in this model. Thus P4 permits each individual configuration to have a different origin *and* a different set of weights. It is just as well that Procrustes' imagination was not fired by modern MDS!

Psychologically, the vector models have a certain appeal, since they allow different *categorisations* of stimuli to be related by regular transformations, and allow for such processes as over-compensation—in these models, the Maoist and the Stalinist can in effect share the same political map whilst consigning each other to the fascist camp by means of a single parameter, the negative vector weight!

These increasingly complex transformations bring better fit, but at a cost. In its more complex models PINDIS becomes prolific in its use of parameters, and many users (especially statisticians) are rightly wary of the degrees of freedom consumed. The parsimony of Occam's razor appears not simply to be blunted but to be thrown away with abandon. Assuming that the number of stimuli considerably exceeds the number of dimensions, then the models assume a natural hierarchy defined by the number of parameters (cf. Lingoes and Borg 1978, p. 495).

Fewest free parameters	*Model*	*Parameters per individual configuration*
	P0	(No parameters) (only admissible transformations)
	P1	r dimensional weights
	P2	r dimensional weights and $\binom{r}{2}$ rotation coefficients
	P3	p vector weights
	P4	p vector weights and translation vector of r elements
Most parameters	(P5	r dimension weights and p vector weights)

When deciding which PINDIS model is most appropriate, one will necessarily be trading off the increase in goodness of fit (or explained variance) against the increase in the number of fitting parameters. There are no reliable statistical ways for deciding whether the trade-off is worth it, so the assessment of which PINDIS model is best will always retain a strong subjective element.*

7.4.2 The distance models (P1 and P2)

The two forms of distance (or, more strictly, dimensional) model in PINDIS examine the extent to which a given configuration can be better fitted to the centroid

*Since this was written, Langeheine (1980) has produced the results of his simulation studies of the PINDIS models, which provide expected fit measures and other statistics as a guide to help the user in deciding on the appropriate model. The use of these approximate norms is strongly recommended.

configuration by differential weighting of the axes (P1) or by differential rotation followed by individual weighting in the case of P2. It should be remembered that in neither case are the original relative distances preserved and in this sense the models involve 'inadmissible transformations'.

7.4.2.1 P1, the weighted distance model (with fixed dimensional orientation)

The specification of this model is formally identical to INDSCAL:

$$\delta_{jk}^{(i)} = d(x_j, x_k) = \sqrt{\sum w_a^{(i)}(z_{ja} - z_{ka})^2}$$

i.e. the Euclidean distance between points x_j and x_k in the original configuration $\mathbf{X}_i$ is assumed to be (a similarity transformation of) the distance between points z_j and z_k in the centroid configuration, after each dimension has been differentially weighted. A detailed comparison of P1 and the INDSCAL model is contained in Borg and Lingoes (1978).

The chief advantage which P1 has over INDSCAL is that in the PINDIS hierarchy it is possible to investigate first how much variation can be accounted for in a given set of data by legitimate similarity transformations (i.e. the Procrustes rotation of model P0) *before* having recourse to individual weights. Only if the improvement in explained variance between P0 and P1 is substantial is it worth proceeding to a more complex distance model. P1 differs from INDSCAL in two other significant ways:

(i)   in the manner of estimating the group space, and
(ii) in the interpretation of the subject space weights.

In INDSCAL, the group space co-ordinates and subject weights are determined simultaneously, whereas in P1 the centroid space is determined first, then put into optimal orientation for dimensional weighting. Only then are the subject weights calculated as a separate operation. This difference has the effect of improving the properties of the subject space. In particular:

> The squared length of the vector drawn from the origin to the subject space point corresponds *exactly* in the P1 model of PINDIS to the variation in the subject's data (individual configuration) explained by the model and is independent of any orthogonality properties of the 'group space' or centroid configuration. (This, it will be recalled, holds only 'approximately' in INDSCAL, depending on the correlation between the dimensions.) It also means that in P1 it is possible to estimate the contribution of each dimension to the total communality, if the user so desires.
>
> The separation of subject vectors in the subject space correctly represents the correlation of the respective configurations in P1. (This holds only approximately in INDSCAL.)
>
> P1 can accommodate negative weights, which can be interpreted as reversed or reflected dimensions.

In actual practice, the application of P1 and INDSCAL to the same data (after preliminary scaling in the latter case) will lead to very similar results in terms of the centroid configuration, but there will often be significant differences in the subject space. This is well illustrated in the Borg and Lingoes 1978 paper.

An example involving the P1 model is given in 7.4.5.

7.4.2.2 *P2, the idiosyncratically rotated and weighted distance model*

This model allows individual configurations ('subjects') to differ both in the frame of reference which they adopt (i.e. in the dimensions they choose as significant, so long as they are orthogonal to each other) and in the weights they attach to them. This is also a dimensional model, but one where an idiosyncratic rotation of the axes of the centroid space occurs before the application of weights. The individual differences in rotation could be trivial or substantial. In most cases using PINDIS, it will be found that a slightly different individual orientation of axes will fit an individual configuration somewhat better than the averaged orientation provided by P1. (Indeed, in the PINDIS computational procedure, the individual orientations of axes are estimated first and the orientation for P1 is then averaged from these.)

On the other hand, some subjects may employ a rotation which is clearly different from others. In psychological terms, this often means that some *combination* of the initial P1 dimensions or properties are more salient than the original ones. If, for instance, it turns out that in the judgment of the Irish political candidates two main consensual dimensions are a pro/anti-British and a left/right dimension, it may well be that for a significant fraction of the population a single left-wing, anti-British *vs* right-wing, pro-British dimension is virtually the only one that matters, with small residual variation on the other dimension. In terms of the P2 model, this would mean that for such people a rotation of about 45° with a large weight on the new dimension I and a small weight on dimension II would provide a better frame of reference than the P1 centroid configuration.

The P2 model is very similar to the Carroll-Chang IDIOSCAL model (the acronym stands for Individual Differences In Orientation SCALing) to a variant of Tucker's Three-mode Scaling and to Harshman's (1972) PARAFAC-2. These are discussed in Carroll and Wish (1973, p. 440 et seq.) and the precise relationship between IDIOSCAL and INDSCAL is discussed in Borg (1979, p. 635 et seq.). The P2 and IDIOSCAL models resemble each other in much the same ways as P1 and INDSCAL do, in particular in having separate *vs* simultaneous estimation of the group space and subject parameters, clearer interpretation of the subject parameters in P2, and estimation from the original data (IDIOSCAL) as opposed to ready-scaled data (P2). These characteristics are discussed in Borg and Lingoes (1978) and Borg (1980), and both include a detailed mathematical treatment and a number of illustrative examples.

In many ways the P2/IDIOSCAL model is very appealing, but it is a rather complex and relatively ill-understood model, and one for which there are not as yet any very compelling empirical examples. Users are once again cautioned to proceed with care.

7.4.3 *The perspective (vector weighting) models (P3 and P4)*

The procedure involved in the perspective models is the construction of a vector from the origin to each stimulus point of the centroid configuration, and the weighting of these vectors differentially for each configuration in order to get it into

VECTOR WEIGHT:	Direction	
	POSITIVE sign (same direction)	NEGATIVE sign (opposite direction)
Size		
LARGER (further out)	Point moves further out in same direction	Point moves a greater distance but away from origin
UNIT (same length)	*Point in same position as in Centroid Configuration*	Point is same distance from origin, but in opposite direction compared to Centroid
SMALLER (closer in)	Point moves closer towards origin	Point moves a smaller distance but away from origin

Table 7.3   *Effect of size and direction on point relocation*

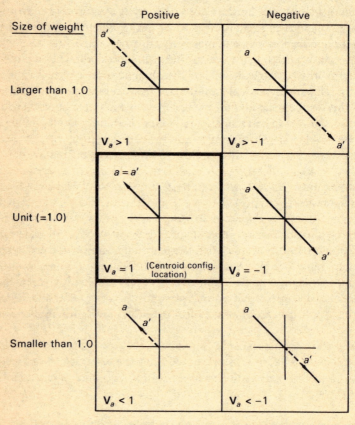

VECTOR WEIGHT (V_a)

Sign of weight

Figure 7.7   *Effects of size and sign of vector weight (V_a) on point relocation*

Note: *a* denotes location of a point in the centroid configuration; *a'* denotes relocation after applying vector weight (V_a).

better fit with the centroid. Taking the original centroid vectors as the unit, each individual vector weight may be smaller, the same or larger, and it may be positive or negative in sign. These differences and their effects are summarised in Table 7.3 and illustrated in Figure 7.7. Note that whatever value the vector weight has, it can only relocate a point in the same or in the opposite direction.

7.4.3.1   P3, the weighted vector model (fixed origin)

In P3 the main focus of interest is obviously on the set of vector weights which consistently transforms the centroid configuration into as close an approximation to this individual configuration as possible. What should be looked for in comparing individual sets of vector weights? For any configuration, the closer these weights are to $+1$, the more that individual configuration resembles the centroid and the less useful the vector model is. Within an individual set of weights interest will normally centre upon which are largest and/or which have negative sign, since these imply the greatest relocation compared to the centroid. It can often happen, for instance, that apart from one or two points, the weights are all close to $+1$, indicating that the significant differences are concentrated in a few points, but that the remaining structure of the configuration closely resembles the centroid. (This incidentally, could never be detected using a dimensional model where all point co-ordinates are *ex hypothesi* systematically weighted.)

In P3 the vector weights are comparable across individual configurations (this is not true of P4) and this provides a second important type of comparison. Presumably, the stimulus points where vector weights vary most from configuration to configuration are the ones which are least stable in the configuration and could be removed from analysis, or alternatively could be given more detailed study. It often happens that the variation in weights for a given stimulus vector is higher in some individual configuration than in others, which suggests that variation is concentrated in a particular area or substructure of the configuration. Once again such a difference cannot be detected using the dimensional models.

In many applications, the simple perspective model P3 shows a dramatic increase in explained variation compared to the basic model (P0), and to P1*, and provides a considerable degree of detail for analysis. The P3 model is further illustrated in Figure 7.8. Here the eight crimes configuration of Figure 7.4b is taken as the centroid configuration. Two sets of vector weights are then applied to produce the 'private spaces' of (b) and (c) in Figure 7.8. As in the INDSCAL and P1 models, the overall shape of the configuration is changed, but in P3 the local structure is also changed, as any cluster analysis would dramatically show!

The main differences between (b) and (c) in their *pattern* of weighting are concentrated in the location of Rape (2) and Libel (5). (b) isolates Rape further from assault (moving it away in a south-westerly direction from all the other points) and (c) projects Rape into the opposite direction entirely, to join receiving as its nearest

---

*Such a result must be treated with some caution, for P3 allows one parameter to be estimated for each stimulus, whereas P1 simply allows one for each dimension. Usually the number of stimuli considerably exceeds the number of dimensions, hence one would expect a better fit for P3 compared to P1. The question is just how dramatic an increase is needed before deciding that it is not simply due to the additional degrees of freedom. See Langeheine (1980) for information relevant to this decision.

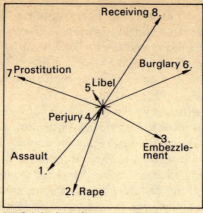

(a) Original configuration (Fig 7.4 b)

PRIVATE SPACES (Fixed origin; dotted vector denotes negative weight)

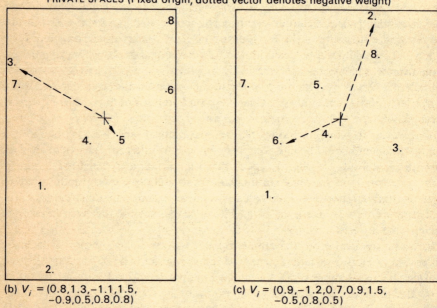

(b) $V_i = (0.8, 1.3, -1.1, 1.5, -0.9, 0.5, 0.8, 0.8)$

(c) $V_i = (0.9, -1.2, 0.7, 0.9, 1.5, -0.5, 0.8, 0.5)$

Figure 7.8 *P3: perspective model (fixed origin)*

neighbour. However, whilst the size of the weight for libel differs most, its effect is not so marked since libel is located fairly close to the origin in the first place and the composite effect is less dramatic. Nonetheless, (b) now locates libel more in the direction of burglary whereas (c) moves it somewhat closer to prostitution.

This example illustrates rather well the point that, when interpreting the P3 model, it is not simply the size and pattern of vector weights that are relevant but also the multiplicative effect upon the original length of the vector. A massive weight on a point located close to the origin can often move a point a very small distance, whereas it only needs a fairly small weight to move a peripheral point yet further away.

With some justice, the P3 model has been hailed as the major innovation introduced into MDS by PINDIS. It certainly provides a powerful and subtle form of analysis of individual differences and often gives insight into the detail about the source of variation in configurations.

7.4.3.2 *P4, the idiosyncratically translated and weighted vector model*

In the words of all good detective stories, 'a little thought should convince the reader' that vectors drawn from different origins will alter the pattern and shape of the configuration. The facts of the matter are illustrated in Figure 7.9. Here the centroid configuration consists of three stimulus points (labelled 1, 2 and 3) which form an equilateral triangle centred upon the origin $(0, 0)$. Initially, the vector lengths drawn from the origin of the centroid configuration are all unity. Suppose now that three individuals all happen to employ the *same* vector weighting, namely $(\frac{1}{2}, -1, 1\frac{1}{2})$. (In the normal way, P4 model individual vector weights can, and will, differ. Making them identical just simplifies the example.)

From the perspective *located at the origin* (labelled A), the original equilateral triangle will be deformed by the weights into the $(1^a, 2^a, 3^a)$ triangle joined by the unbroken line. But from the perspective of B at $(-2, 2)$, the configuration $(1^b, 2^b, 3^b)$, denoted by the dotted line, looks quite different, and it looks different again from C's perspective at $(-1, 2)$. Convince yourself that the differences between A's, B's and C's triangles arise purely from the fact that they have different origins. (The configurations would be different again if they did not share the same vector weights.)

For obvious reasons, P4 is often called the 'points of view' model (though this term was originally used by Tucker to refer to a quite different model.) In P4, it is the idiosyncratic origins that can be directly compared and that form the main focus of attention. For instance, the question of whether two subgroups of subjects differ significantly in their perspectives can be readily investigated. But the individual vector weights cannot be directly compared *except* in conjunction with the idiosyncratic shift of origin, which would mean constructing a new set of vectors all emanating from the same origin.

There have been few studies to date which have used this model, and only one where the increase in explained variations is impressive (Lingoes 1977).

7.4.4 *Variants and options within* PINDIS

As currently programmed, the user may also use PINDIS in an 'external' manner by inputting a hypothesis configuration instead of calculating the centroid. If so requested, this configuration will *not* be differentially rotated (for the dimensional models) and the origin will *not* be translated (for the vector models). The net effect is to suppress P2 and P4 respectively; by a separate user-controlled parameter, P5 can also be suppressed.

The external use of PINDIS is most appropriate when a target configuration is being used as a fixed reference point for other configurations (as in replications and confirmatory studies) and where a known (or hypothesised) structure such as physical properties or a geographical map underlies the data. Note that a hypothesis matrix can be input (to replace the centroid configuration) without also requiring that it be fixed.

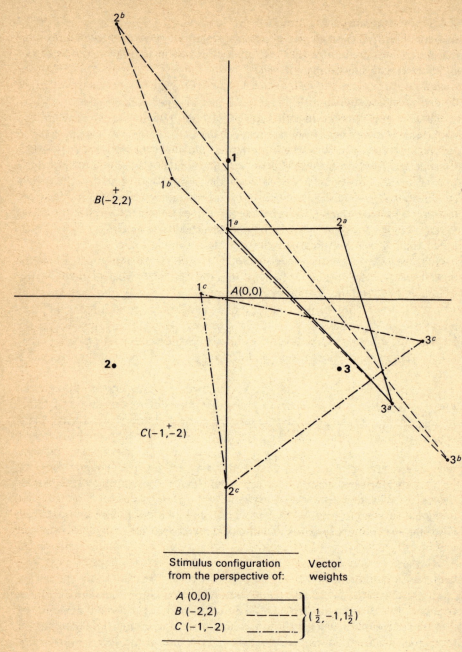

Figure 7.9   *P4: perspective model (idiosyncratic origins)*

Finally, the user may choose to have a specific stimulus location as the origin of the centroid configuration for the vector models. This is especially useful if the configurations being compared have a radex structure, with some stimulus item forming the natural origin. Details of the program parameters used to implement these options are given in MDS(X) *User's Manual* documentation for PINDIS.

7.4.5 *An application of* PINDIS

A range of published examples of the application of PINDIS is included in Table 7.4. It is interesting to note that the 'optimal' PINDIS model often turns out to be either the simple dimensional model (P1) or the simple vector model (P3). Only the somewhat atypical second example of Lingoes (1977) gives strong support to the double weighting model.

These examples also show that PINDIS can be used in an exploratory or a confirmatory mode, or in combination of both.

A further example, also drawn from the occupational cognition study, illustrates these points. A set of 48 subjects of varying status and occupational backgrounds were asked to rate or rank 16 occupational titles on three criteria frequently employed by sociologists to serve as forms of 'prestige', to which were added two factual criteria concerned with estimated average income and how much the subject thought she knew about the occupations concerned.

The data were scaled internally using MDPREF with the following values of an overall goodness of fit measure (linear root mean square correlations) of the data to the 3-D MDPREF solution. (Data, solutions and further details are contained in Coxon and Jones 1979, pp. 86–105):

Criterion		Goodness of fit RMS, 3-D MDPREF solution with data
1. *Social standing*	(own opinion of general standing in the community)	0.90
2. *Prestige and rewards*	(an occupation ought to get)	0.89
3. *Social usefulness*		0.84
4. *Monthly earnings*	(estimated by subject)	0.87
5. *Cognitive distances*	(how much subject knows about what a job involves)	0.73

We were interested in knowing how these MDPREF solutions differed among themselves, and also how they compared to the INDSCAL group space configuration (obtained by separate estimation of similarity and discussed earlier under 4.6 and illustrated in Figure 4.5). The differences among the MDPREF configurations were first investigated by using a straightforward PINDIS analysis. Their similarity to the 3-D INDSCAL group space was then analysed by using it as a fixed hypothesis configuration in a second run.

The result of the first analysis is summarised in Table 7.5, and the centroid configuration is given in Figure 7.10. The resemblance between this and the INDSCAL configuration of Figure 4.5 is, at least at first sight, very marked indeed. In terms of similarities between the configurations (Table 7.5a), there is quite good internal agreement: on average 69 per cent of variation between the configurations is attributable to admissible transformations, with the two status criteria (1 and 2) being especially well fit and estimated earnings (4) particularly badly fit. Neither of the dimensional models improves this fit in any way at all—a mere 2 per cent increase is involved. On the other hand the vector models do rather better, with 17 per cent improvement for P3 and 24 per cent improvement for P4. Once again, the earnings configuration fits relatively poorly, but its fit is markedly improved by allowing it to have an idiosyncratic origin.

Reference	'Subjects' Configurations	Stimuli	PINDIS models P0 P1 P2 P3 P4 P5	Fit (r^2) P0	Fit (r^2) 'Optimal'	Comments
Lingoes 1977	(i) 2 species-related, but apparently dissimilar, fish structures.	29 points defining fish outline	√ √ — √ — √	0.72	P5:0.96	Dimensional weighting does better than vector despite fewer parameters. One of the few instances where P5 produces markedly better fit.
	(ii) 3 biological (genetic) 'maps' measuring dermaglyphic, anthropometric and genetic characteristics of the Indian population.	7 Latin American villages	√ √ √ √ — √	0.63	P3:0.90	Interesting use of target (geographical) configuration which is *not* kept fixed. Clear superiority of P3 over P1 suggests interesting conclusions (q.v.).
Borg 1980 (i) *data* Lingoes et al. (1979)	14 individual subjects (SSAR-1)	6 Political Party × 6 'Closeness' combinations	√ √ √ √ — √	0.77	P3:0.88	Not a marked improvement.
(ii) *data* Green and Rao (1972)	41 individual subjects (MDSCAL)	15 breakfast foods	√ √ √ √ — —	0.72	P3:0.85	Lot of individual variability.
(iii) *data* Lingoes et al. (1979)	10 normal (N) and 4 colour-dificient (CD) subjects (Classic Scaling)	10 colour tiles	√ √ √ √ — —	N:0.98 CD:0.74	(P0) P1:0.88	Z of normals used as fixed target for assessing colour-deficients. As expected, simple dimensional weighting of colour-circle is best model.
Maimon et al. 1980	4 groups of managerial graduates and supervisors. (SSA-1, 2D)	9 items (facet-design) type × area of skills	√ √ √ √ — √	0.71	P3:0.90	High variability on all other models.
Coxon and Jones 1980	6 individual deviant cases from occupational cognition project (MINISSA, 3D)	16 occupations	√ √ — √ — —	0.29	P3:0.70	Used fixed target (INDSCAL) configuration.

Table 7.4 *Applications of* PINDIS

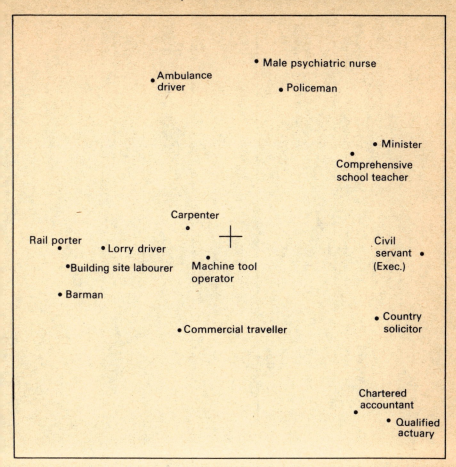

Figure 7.10   *First 2 dimensions of PINDIS centroid configuration derived from 5 MDPREF configurations*

Concentrating now on P3, where are the discrepancies located? An analysis of the vector weights indicates unequivocally that the first three status criteria and the last two ('cognitive') criteria agree very considerably among themselves, and contrast with each other in terms of the relative size and pattern of the weights. Most 'unscrambling' is done with respect to the machine tool operator (the prestige configuration has a vector weight of 2.43, whilst the cognitive distance configuration has one of the few negative weights of -0.17 for this occupation), and to a lesser extent the civil servant (the earnings configuration attaches a weight four or five times that of the other criteria).

In the second run, the configurations were related to the INDSCAL 3-D configuration—which, as we have commented, the PINDIS centroid configuration seems to resemble closely—the degree of fit drops dramatically (see Table 7.5b). Clearly there *are* very considerable differences with respect to the INDSCAL configuration and we have been over-impressed by their surface resemblance. Nor is the fit improved by the dimensional model. Since the INDSCAL configuration is fixed,

Communalities (r^2) for PINDIS Transformations

Configuration	P0 Basic	P1 Dimensional weighting	P2 Dim. weighting and rotation	P3 Vector weighting	P4 Vector weighting and translations	P5 Double weighting
1 Social Standing	0.83	0.83	0.83	0.94	0.99	0.88
2 Prestige and Rewards	0.83	0.84	0.84	0.96	0.92	0.90
3 Social Usefulness	0.77	0.78	0.78	0.86	0.92	0.83
4 Monthly Earnings	0.40	0.42	0.44	0.68	0.89	0.81
5 Cognitive Distance	0.62	0.64	0.64	0.82	0.93	0.73
Average	0.69	0.70	0.71	0.86	0.93	0.83

(a) Full analysis (all simple models, no hypothesised configurations)

	P0	P1	P2	P3	P4	P5
1 Social Standing	0.26	0.27	—	1.00	—	0.64
2 Prestige and Rewards	0.16	0.17	—	0.79	—	0.65
3 Social Usefulness	0.19	0.20	—	0.55	—	0.47
4 Monthly Earnings	0.12	0.12	—	0.40	—	0.57
5 Cognitive Distance	0.25	0.26	—	0.62	—	0.59
Average	0.19	0.20	—	0.67	—	0.58

(b) Analysis with INDSCAL 3-D configuration as hypothesis

Table 7.5 PINDIS *analysis of 5 occupational criteria configurations*

the analysis excluded rotation of axes and translation of origin, so the parameters of P2 and P4 were not estimated. Once again, the simple vector model does a dramatic job of improvement, with the social standing configuration now fitting perfectly, and the earnings configuration still being rather poorly fit.

Such a finding is substantively important: it shows for instance that we can conclude that data obtained from judgments of 'general standing' of an occupation give rise to a conclusion that is a systematic transformation of the INDSCAL structure obtained from judgment about 'overall similarity'. At the very least, such a finding strongly contests the received opinion that the latter is a cognitive criterion and is generically distinct from the evaluative status criteria.*

7.5 Preference Mapping (PREFMAP I and II)

The Carroll-Chang (Carroll 1972, p. 114 et seq.) preference mapping models form a hierarchy of models, akin in many ways to PINDIS. They differ principally in that PREFMAP (PM) is designed primarily for *external* scaling (where the user provides a stimulus configuration) and the input data consist of a rectangular matrix of ratings or rankings of p stimuli given by N subjects. The purpose is to map *each subject* into the stimulus space in the form of an ideal point (PM Phases I–III) or as a vector (PM Phase IV) according to a hierarchy of increasingly complex models. (The transformation may be linear/metric or quasi non-metric/ordinal according to the user's choice.) We have already encountered earlier in this book the two simplest models: the *simple distance model* (Phase III) (5.3.3.1 and 6.2.4) and the *vector model* (Phase IV) (5.3.2 and 6.2.2) and so the focus here will be on the more complex models—the rotated and weighted distance model (Phase I, akin to P2 in PINDIS) and the weighted distance model (Phase II, akin to P1 and INDSCAL). But it will be helpful to begin by summarising the full range of the PREFMAP models.

7.5.1 The PREFMAP *hierarchy of models*

The basic notion of all the models in PREFMAP is that an individual's preference ranking is a function of the distance separating her point of maximum preference (ideal point) and the stimulus points. It is the way in which the distances are defined which differentiates the models.† (PREFMAP can also be seen as extending and generalising Coombs' unfolding model, and as showing that the vector model of preference is a special case of unfolding.)

Briefly, the hierarchy of models is as follows:

I *General Unfolding Model (PM1)*

This is the most general model. Each subject is viewed as having a specific, most-

---

*When a two-dimensional PINDIS analysis is performed, the results are much the same, but even more marked. Admissible transformations account for 75% of variation in the 'internal' PINDIS analysis and for 14% variation when related to the 'external' INDSCAL configuration. In neither case do the dimensional models add more than a derisory amount (2% in both cases), but the increase due to simple vector weighting is more impressive (18%, and 57% in the external case). But the 'unscrambling' is different in detail; the machine tool operator is still relocated more than any other occupation, but the actuary and accountant are also seen to be relatively unstable in their positioning.

†Technical details of the models are given in Carroll (1972) and in Coxon and Jones (1979, p. 106 et seq.), and a detailed exposition of the computational procedure and output details is given in van Schuur (1977).

preferred ideal point in the space, rotating the axes of the similarity space to his own reference dimensions, and then attaching an evaluative weight to each of them.

II   *Weighted Unfolding Model (PM2)*
Subjects are assumed to have an ideal point and to share the same set of reference dimensions (i.e. no individual rotation), but to evaluate or weight the dimensions differentially.

III   *Simple Unfolding Model (PM3)*
Subjects are assumed to share the same set of reference dimensions *and* to give the same weighting to them. However, subjects do still differ in terms of where their ideal points are located in the space. This is the external analysis analogue to Coombs' unfolding analysis.

IV   *The Vector Model (PM4)*
Carroll (1972b) has shown this to be a special case of Coombs' unfolding, when an ideal point is located far from the origin of the space.* It is for this reason that it features in the hierarchy, since all the other models are straightforward *distance* models. In the vector model, by contrast, subjects are represented as a vector (or line directed toward the region of greatest preference), and a preference ranking is interpreted as the order of the projections of the stimuli points on this line. (The internal analogue of this model is MDPREF.)

One particularly valuable aspect of the hierarchical nature of the PREFMAP models is that it is possible to test whether a more complex model explains significantly more variation in the data than a simpler one. In this way, the model which makes the most parsimonious set of assumptions can be chosen. Moreover, as in PINDIS, there is no reason why one particular level or model should be thought to apply to all subjects—it might well be, for instance, that whilst the simple unfolding model applied to most subjects, the data of the remaining subjects might be far better explained by assuming that they simply differentially weight the dimensions of the space.

All the PREFMAP models make two basic assumptions:

(a)   *A common similarity space is assumed to apply to all the subjects included in a* PREFMAP *analysis.*   If it turns out that this does not hold (because, for example, a previous INDSCAL analysis of the similarities led to the conclusion that subjects were divided between distinct 'points of view'), then a separate PREFMAP analysis should be run for each subset, using the group space for each subset as the input configuration.

(b)   *An individual's preference values for a set of stimuli are assumed to be linearly (or in the non-metric version, monotonically) related to the distance between her ideal*

---

*Intuitively this can best be seen in terms of isopreference contours (see Figure 7.12). For the basic distance model, all stimuli at a given distance from a subject's ideal point form a circle (isopreference contour) in 2-space, a sphere in 3-space, and so forth. For the vector model all stimuli at a given distance along a subject's vector form a line (projection) in 2-space. If an ideal point is taken further and further out from the origin, the circular isopreference contours come closer and closer to being a line in the vicinity of the stimulus points. See Carroll (1972).

point and the stimulus points. This assumption is only made for the first three models (which are distance models), and is not made for the vector model.

To give these ideas some substance, let us return to the example where a sample of respondents had been asked to assess the general similarity of a set of Irish politicians, and then been asked to rank them in order of personal preference. Let us also assume that the INDSCAL analysis of their similarity data indicated that they had fairly similar perceptions of the politicians and that the two main differentiating axes were left/right orientation and Republican/Unionist. The INDSCAL group space configuration could then be used as the 'independently established' cognitive space for input PREFMAP, and the focus of interest would now shift to explaining differences in the preference rankings of politicians which the respondents gave.

The simplest *vector model* (Phase IV) of PREFMAP assumes (as in MDPREF) that the subjects collapse the multidimensional stimulus space into one dimension (or line) representing the order of preference. Individual differences in preference are expressed by the differing directions which the vectors have in the common space. In this example, it is conceivable that some subjects simply prefer politicians who combine radicalness with support for Irish Republicanism (or conservatism with union with Britain), whilst others evaluate a politician solely in terms of how left-wing she is, or how much she supports Irish Republicanism.

The *simple unfolding (distance) model* (Phase III) assumes, by contrast, that each subject has one most preferred point in the group space, and that this serves as a reference point for evaluating the politicians, according to how close they are to her ideal point. The *weighted unfolding model* (Phase II) assumes that subjects differ considerably in the value they attach to the dimensions of the space—figuratively, that they pay attention to how highly they value or weight each dimension before they decide how close a politician is to their ideal. The effect of this is to 'pull' a politician closer to the subject's ideal point than would be the case in the simple unfolding model, if the politician occupies a high position on a dimension which the subject highly values.

The *general unfolding model* (Phase I) drops the assumption that subjects' evaluations refer to the same fixed set of dimensions. Instead, it allows them to structure the space as they wish by providing their own reference axes (so long as the dimensions they choose are not correlated) and *then* it allows them differentially to evaluate these 'private dimensions'. The effect of this is that subjects can be allowed to place high evaluation on *combinations* of the original dimensions. For instance, if the original axes were rotated anticlockwise through 45°, subjects who chiefly prefer politicians who combine socialism *and* republicanism (but who still differentiate socialist-Unionists from conservative-Republicans) could easily be accommodated.

Turning now to the hypothetical results of a PREFMAP analysis, it might turn out that, on average, the simple unfolding model (II) held best—that is, most subjects evaluated the politicians in terms of highly salient characteristics (dimensions) which were evaluated in the same way, and only differed substantially in what their positions were on the dimensions. But the data of a minority of subjects might be much better explained by assuming, *in addition*, that they valued highly politicians who were socialist-Republicans, cared not at all for conservative-Unionists but still made some (but relatively little) differentiation between socialist-Unionists and conservative-Republicans.

7.5.1.1   PREFMAP *Phases I and II*
Phase I: General unfolding model
Subjects are permitted

    (i)   to rotate the reference dimensions of the space, and
    (ii) *then* to weight them differentially.

In Carroll's terms:

> We allow distinct individuals additional freedom in choosing a set of 'reference axes' . . . and then to weight differentially the dimensions defined by this rotated reference frame, in addition to being permitted an idiosyncratic ideal point.
>
> (Carroll 1972, p. 120)

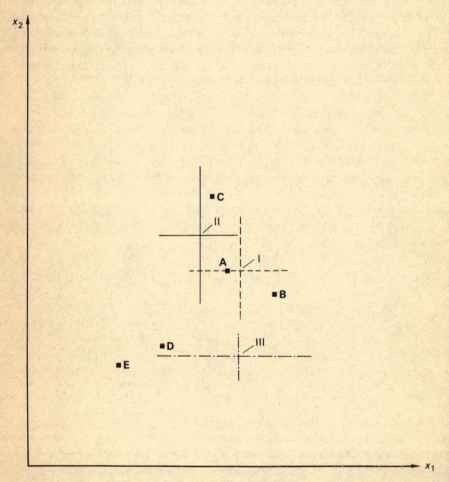

Figure 7.11   *PREFMAP, Phase II (weighted unfolding): subject ideal points and weighted dimensions*

A subject is assumed to apply his own orthogonal rotation to both the stimulus and ideal point co-ordinates, and *then* weight the rotated dimensions. If x_{ja}^* represents the *transformed* stimulus co-ordinates, and y_{ia}^* the transformed ideal points, then

$$s_{ij} = F_i(d_{ij}^2)$$

where $\quad d_{ij}^2 = \sum_a w_{ia}(y_{ia}^* - x_{ja}^*)^2,$

i.e. a Euclidean distance in an individually-rotated and weighted 'private space'.

Phase II: Weighted unfolding model
This is similar in many ways to the INDSCAL and PINDIS P1 model except that in this *preference* model, the weights are interpreted as reflecting the subjects's *evaluation* of the dimension when making an overall preference judgment. To continue the Irish example—two subjects might well entirely agree about, say, the left-right orientation of the politicians. To one subject this may override all other considerations when it came to choosing a candidate, whilst to another it might be considered entirely irrelevant compared to the politician's position on Republicanism.

In this model, a subject is assumed to apply an evaluative weight w_{ia} to each dimension, so that

$$s_{ij} = F_i(d_{ij}^2)$$

where now $d_{ij}^2 = \sum w_{ia}(y_{ia} - x_{ja})^2.$

i.e. a Euclidean distance in a weighted 'private space'.

The weighted unfolding model (PM2) is illustrated in Figure 7.11 (and see also Figure 7.12). In Figure 7.11 the joint space includes five stimulus points (A to E) and the location of the ideal points of three subjects (I, II and III). In this model, the subjects differ in terms of the location of their ideal points (akin to PINDIS idiosyncratic perspective model (P4)†), and they also attach different evaluative weight or salience to the dimensions. In Figure 7.11 the differential weights are denoted in the form of the arms of a cross. In the case of I, the weights for the dimensions are equal; for II the weight of dimension II is greater than that for dimension I, and for III the reverse is true. Note that in this model the individual axes are all oriented in the same direction, parallel to the reference axes and are therefore directly comparable. 'Private spaces' for each individual can be produced, if so desired, by stretching and shrinking the reference axes. A more convenient representation is illustrated in Figure 7.12.

7.5.2 The a priori *stimulus space*
Since the PREFMAP models are designed to be external in form, the user must normally supply an *a priori* configuration, and similar issues arise as in the case of PINDIS: what may reasonably be used as an *a priori* configuration? How, if at all,

---

†Compared to the PINDIS hierarchy, PM2 is a hybrid model. PM2 is similar to P4 in the sense of allowing differing points of view, but is similar to P1 in allowing differential weighting of axes.

does the configuration change at different levels?

The source of the *a priori* configuration can be considered under three heads: (i) a previous scaling, (ii) a theoretical or rational configuration and (iii) an 'internally-generated' configuration.

A previous scaling

The configuration may be the result of a previous MDS scaling analysis of similarities data for the same set of stimuli, probably obtained from the same subjects that provided the preference data. This is probably the most common instance in social science applications.

When replicating a study it can be useful to see how well the data from one's own study will fit the configuration obtained by the original investigator.

A theoretical or rational configuration

By a 'rational configuration' is meant one which incorporates the actual characteristics of the stimuli as dimensions of the *a priori* space. This occurs typically in psychophysical applications and in cases where the stimuli are well-defined compositions. Examples include the now infamous 'hypothetical cups of tea' data (collected by Wish and analysed in the Carroll 1972 paper) whose defining characteristics were the *hotness* and the *sweetness* of the brew, and in the Delbeke-Bollen family composition data (cf. Coxon 1974 and section 6.2.2) whose characteristics were the *number of sons* and *number of daughters* making up a family. Other instances might include a facet analysis of the stimuli, the geographical location of stimuli, or some other theoretically expected or physically underlying configuration.

Internally generated configuration

The option also exists in PREFMAP for a stimulus configuration to be constructed from the *same* preference rankings data as are used in the analysis. (It is implemented in the MDS(X) version using the INITIAL parameter.) Such a 'quasi-internal' configuration is constructed by forming the minor (stimulus × stimulus) product moment matrix from the data, (which may have been transformed in some user-specified way*), and then performing a classic basic scaling (see Appendix A5.2) to yield the configuration.

In the MDS(X) version of PREFMAP, the user also has the freedom to keep the stimulus space configuration fixed throughout all phases of analysis, or to allow the program to modify it to obtain a better fit to the data by the use of the KEEP parameter. If the first option is chosen, the original stimulus configuration remains unchanged throughout the analysis. In the second case, the form of modification is slightly different, depending upon the level of the model (phase) at which the user starts analysis (see Carroll and Chang 1967, p. 10 et seq. and Carroll 1972, p. 134 et seq.).

---

*Options exist within the INITIAL parameter for double centring the product matrix (removing row and column means and replacing the overall mean); removing row means; standardising rows, and for removing row and column effects (see *User Manual*, PREFMAP Report, 2.3.1). If the first option is chosen and the data then analysed under the linear version of PM3, the resulting analysis turns out to be equivalent to an internal metric unfolding analysis. On this and related issues, see Carroll 1970, 1971, Carroll and Chang 1970, and Schonemann 1970.

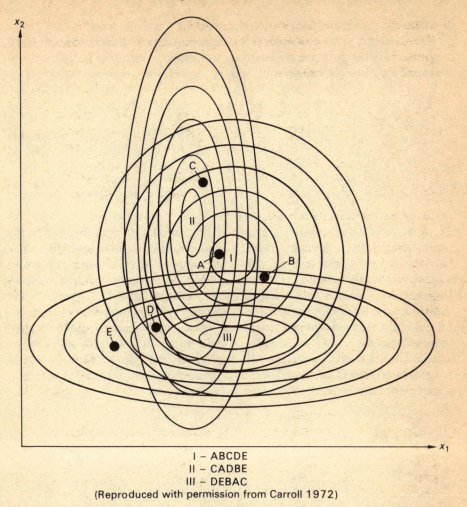

I – ABCDE
II – CADBE
III – DEBAC
(Reproduced with permission from Carroll 1972)

Figure 7.12   *Weighted unfolding model (PREFMAP Phase II): isopreference contours for three individuals*

7.5.3   *Preference and 'anti preference'*

In the distance (unfolding) model of preference, it is assumed that a subject's preference is single peaked and symmetric—that she has one point of maximum preference in the stimulus space and that preference decreases systematically in all directions (along each dimension). One way of representing this is to define a set of contours (at arbitrary but equal intervals) centred upon the ideal point. Each contour represents points at an equal distance from the subject's ideal point, and hence all points will be equally preferred. In preference analysis, these contours are referred to as 'isopreference contours'. In the case of the simple distance model (PM3), where each dimension is equally weighted, these contours will in the two-dimensional case be circular, and in the weighted distance model (PM2) they will be ellipses the length of whose axes will equal the value of the evaluative weight. This is illustrated in Figure 7.12, which presents the isopreference contours for the three subjects of Figure 7.11. Note that subject I has equal weights (and hence circular preference contours) whereas the others have unequal, and hence elliptical contours. Once the preference contours are drawn in, it becomes straightforward to reproduce a subject's preference ranking by following the relevant contours. Thus, for subject II, C is within the second ellipse from the ideal point, A within the third, D within the fifth, B just beyond the sixth and E further beyond it—yielding the I-scale: CADBE.

In the case of the rotated and weighted model (PM1) the ellipses are oriented in different directions, but the same logic applies. In the case of the vector model (PM4) the isopreference contours are straight lines, rather than circles or ellipses, since all points which project onto the same point along the preference vector, wherever they be located in the space, will be equally preferred.

The PREFMAP models specifically allow for negative evaluative weights. How are they to be interpreted? Carroll argues that it is simply a matter of changing the form of the preference function in the distance models, from being single-*peaked* and symmetric to having a single valley (one single point of *minimum* preference) and symmetric. In this case, the contours will decrease towards the subject's 'pessimal', 'anti-ideal' or least preferred point. (Perhaps the simplest example is the preference for the heat of liquids: many people dislike lukewarm tea most of all, but increasingly prefer both hot and iced tea.) It will sometimes happen that a subject's point will include a mixture of negative and positive weights; this situation has been explored in Carroll (1972, pp. 121–3), who shows that in the two-dimensional case the function will be saddle-shaped. Nonetheless, it is often difficult to give meaning to such mixed-sign combinations and some care is needed in their interpretation.*

7.5.4 Goodness of fit

As a nested set of models, it is possible to use analysis of variance statistics as an indicator of how much better one model does than another (or how much more variation one explains) than another. At the end of a PREFMAP run a table is printed containing correlations, F ratios for each subject by each phase/model and between pairs of phases/models, together with root mean squares (RMS) for each phase. The correlations are between the squared distances of the model and the preference data values (or corresponding disparities in the case of the quasi-nonmetric option having been chosen).

The average RMS values usually give an indication of the level of the hierarchy of models which is generally most appropriate, i.e. that level where the marginal increase in RMS values for more complex models is very small. The subject-phase correlations can indicate whether any phase fits a subject's stated preferences, and the extent to which one phase fits better than another. In conjunction, these measures enable the user to decide upon the most appropriate model (phase) of PREFMAP for representing the data, and they also allow subsets of subjects to be allocated to different levels.

Since each model is a special case of the one immediately above it, it is justifiable

---

*Davison (1976) has proposed and recently implemented (Davison 1980) a variant of preference mapping which allows the evaluative weights to be constrained to non-negativity or non-positivity.

to test for goodness of fit of a model to the data by the usual ANOVA procedures, and test for significant increases in variation explained by means of an F-test between models, although the models are clearly not independent.

Most users enter the PREFMAP program at a higher phase/model and drop down to a lower phase. Carroll (1972, p. 135) argues that the solution for the 'average subject' of the highest phase entered should form the basis for the succeeding phases.

Thus, if S-PHASE (1) and E-PHASE (4) are chosen then the full set of models will be applied to the data. This will mean:

(i)   After Phase I (PM1), the rotated reference axes of the average subject form the 'canonical reference frame' for Phase II (PM2).

(ii)   The subject's weights and ideal points are fitted in *this* reference frame in Phase II (PM2), rather than in the original input frame.

(iii)   In a similar manner, the average subject's dimensional weights are applied to the rotated configuration for analysis in Phase III (PM3).

These operations are not always desirable, and the remedy is either to return to the original input configuration at each phase, by setting KEEP (1) or to enter at a lower phase. For instance:

to keep the *orientation* of the original configuration, enter at PM2
to keep the dimensions unweighted, enter at PM3.

The parameters of the model are estimated through quadratic and linear regression (for the metric version) and subsequent monotone regression (for the non-metric version). The former allows statistical testing, on the assumption that the data and fitted values are linearly related, but the latter only permits approximate statistical tests and they should be used with caution.

In many applications of PREFMAP, the main differences seem usually to occur between the simple distance (unfolding, P3) model and the simple vector (P4) models, with few convincing examples of the necessity of invoking more complex models except for particular subjects.

7.5.5   *Uses of* PREFMAP
The PREFMAP program can be used in three principal ways:

(i)   to map individual ratings or rankings into an independently obtained external configuration (external joint mapping).

(ii)   to first estimate the stimulus configuration from the subject's data, and then map the same rankings/ratings into it (quasi-internal mapping).

(iii)   to use PREFMAP to extend two-mode scaling to include additional subjects (extending two-mode solutions).

7.5.5.1   *External joint mapping*
The first is undoubtedly the most common use and the one for which PREFMAP was originally designed. In most cases, the researcher has obtained both pairwise similarities data and preference rankings or ratings from the same subjects. The

similarities data will typically have been scaled by INDSCAL, or averaged over all subjects and scaled by MINISSA, and the resulting stimulus configuration (group stimulus space in the case of INDSCAL) is then input as the external configuration.

7.5.5.2   Quasi-internal mapping
So far there are no published examples of quasi-internal preference mapping. However, as mentioned above (see Carroll and Chang 1971), choice of the metric option FIT (0) with a double-centred configuration generated from the preference data GENERATE (0) provides an optimal solution for an internal *metric* unfolding analysis which, given the frequent instability of the non-metric unfolding analysis (implemented by MINIRSA), may well be a desirable option to choose.

7.5.5.3   Extending two-mode solutions (large data sets)
Users often want to scale data sets where the number of subjects is too large for stated program limits. In these circumstances PREFMAP may be used to effect an increase in the number of stimuli or subjects. A reasonable strategy is to proceed as follows:

(i)   Sample the maximum permitted number of subjects and include the relevant data in a preliminary run of scaling programs to produce a 'core scaling'.

(ii)   Take the output configuration and treat it as an 'external' configuration for PREFMAP analysis.

(iii)   Select the options within PREFMAP which match the model and transformation of the original scaling.

(iv)   Include the remaining data as input 'preference' data for the PREFMAP run, which will then estimate the positioning of the new subject points (or vectors) within the existing configuration.

Although this procedure is most appropriately used for extending the number of subjects for two-way, two-mode scaling models (such as MDPREF and MINIRSA), it may also be used for basic one-mode scaling models (such as SSA and MRSCAL) if the additional data consist, for each new point, of a set of relative distance estimates between the new point and the points already in the fixed configuration. Kruskal (1972b, pp. 5–6) shows how he used essentially this procedure when faced with scaling a data matrix of 10,000 rows (computer malfunctions) by 657 columns (diagnostic tests) to get a configuration of 10,000 points. In essence, subsets of a manageable size were first used to get a rough estimate of the structure and dimensionality of the data. Then a core set of points (in Kruskal's case, 20) is chosen in such a way that they are as well spaced in the configuration as possible. Finally, the remaining points are positioned into the configuration with respect to the core points, by metric or non-metric scaling, as implemented by PREFMAP. This and other ways of scaling large data sets are discussed in Golledge et al. (1981).

Care should be taken to ensure that the transformation is correctly specified (by the parameter FIT) and that the correct model is chosen—S-PHASE (3) for distance models and S-PHASE (4) for vector models.

7.6   Interrelations of INDSCAL, PINDIS and PREFMAP Models
The more complex models discussed in this chapter have a family resemblance.

INDSCAL, PINDIS and PREFMAP each consists of a set of models of increasing complexity, and the type of complexity is also alike, including distance models involving differential dimensional weighting and rotation, and one or more vector models. The relationship can be portrayed as follows:

| | Program/Series | | |
Model	(a) INDSCAL	(b) PINDIS	(c) PREFMAP
Rotated, Weighted Distance	(IDIOSCAL)	P2	PM1
Weighted Distance	INDSCAL	P2	PM2
(Simple Distance)	SSA/MRSCAL	P0	PM3
Vector	—	P3, P4	PM4

The parallelism is not exact, and not all of the programs exist in the MDS(X) package but, arrayed in this way, certain common elements become clear:

(a) The INDSCAL series is the most basic, and take their input from one or more (dis)similarity matrices. The most complex model, IDIOSCAL (IDIosyncratic Orientation SCALing) is a fairly predictable generalisation of INDSCAL where each subject is thought of as carrying out an idiosyncratic rotation of the group stimulus space axes, followed by a weighting of those axes. Unfortunately, as a program it has a number of sub-optimal characteristics, and far from convincing examples of its use have been published. The basic references are Carroll and Wish (1973, pp. 90–2 and 1974, pp. 440–1).

The basic metric and non-metric models, MRSCAL and MINISSA, are obviously special cases of INDSCAL—when each subject has identical sets of weights. There also exists in the original MINI series (and elsewhere) variants of the basic vector ('factor analysis') model, (see, for example, Lingoes et al. 1979, pp. 268–9 (SSA-III) and pp. 307–9 (MINI-NFA)). The widespread availability of (metric) factor analysis programs and of the closely related classic scaling procedure in MRSCAL make its inclusion in the MDS(X) series superfluous. In its non-metric form it has rarely been used, since product moment correlations are (inversely) monotonic with distances (see Appendix A2.1), making it a somewhat redundant model.

(b) The PINDIS series distance models correspond closely to their INDSCAL analogues, except that they take individual configurations as input. The PINDIS series also includes the vector series and the hybrid P5 (double-weighting) model.

(c) The PREFMAP series differ from the other two series in taking rectangular (row-conditional) data as basic input, and representing each row element by an ideal point (distance models) or an ideal vector. Otherwise, the models correspond to their analogues in the other series.

It should also be clear by now that the form of transformation chosen is relatively unimportant compared to the form of the model. Indeed, it is rather ironic that after the 'non-metric revolution' has been accomplished it turns out that in many cases the linear (metric) assumption is a good (and considerably less costly) approximation to the monotonic one. That said, it is as well to err on the side of caution and allow the extensive family of monotonic functions to suggest what more regular transformation might be more appropriate.

APPENDIX A7.1   BASIC GEOMETRIC TRANSFORMATIONS (TWO-DIMENSIONAL)

A.7.1.1   Linear mapping or transformation of a point: *P → P′*

A point is defined by its co-ordinates on each dimension: $P = (x, y)$. If two points are related by a linear equation, then one is said to be a *linear mapping* of the other:

$$\begin{cases} a_1x + b_1y = x' \\ a_2x + b_2y = y' \end{cases}$$

This defines the mapping of point $P = (x, y)$ into point $P' = (x', y')$. Put in another way, the coefficients a and b can be gathered into a *transformation matrix* **T** which, when applied to P, maps it into P':

$$\begin{pmatrix} a_1 b_1 \\ a_2 b_2 \end{pmatrix} \begin{pmatrix} x \\ y \end{pmatrix} = \begin{pmatrix} x' \\ y' \end{pmatrix}$$
$$\text{T} \qquad \text{P} \quad = \quad \text{P}'$$

Example **Q′** is a linear mapping of **Q** (Figure A7.1)

$$\begin{pmatrix} 0.3 & 0.2 \\ 0.1 & 0.4 \end{pmatrix} \begin{pmatrix} 6 \\ 5 \end{pmatrix} = \begin{pmatrix} 2.8 \\ 2.6 \end{pmatrix}$$
$$\text{T} \qquad \text{Q} \quad = \quad \text{Q}'$$

A.7.1.2   Origin and translation

The *origin* of the space is normally defined as the point $(0, 0)$, and all other points are defined by reference to this origin.

It is sometimes useful to move the origin of the space to a new position. This is termed *translation* of the origin (and axes). A particularly common translation is to the centroid—the average co-ordinate or centre of gravity—of a configuration of points. If the origin is moved to (a, b), then the co-ordinates of a point $P = (x, y)$ relative to the *new* origin are given by: $P = (x - a, y - b)$.

In the example below, if the origin is translated to $(3, 4)$ then the point **P** has the co-ordinates $(2, 1)$ in terms of the old origin and has the co-ordinates $(-1, -3)$ in terms of the new origin (Figure A7.2).

Note Translation preserves distances (identically).

Translation does *not* preserve scalar products or angular separation between vectors (see Appendix A2.1, Figure A2.3).

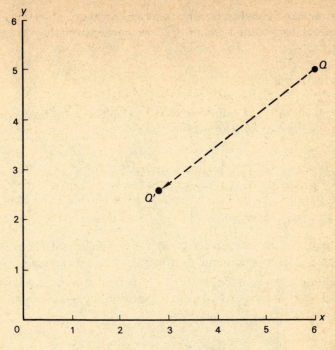

Figure A7.1   *Linear mapping*

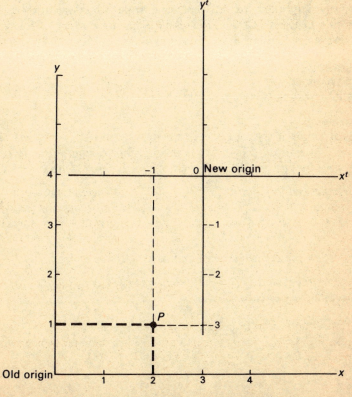

Figure A7.2   *Translation of origin and axes*

A7.1.3 Elementary transformations (null and unit)

Two very basic transformations (which serve as the geometric analogue of zero and unity in algebra) are the *null* and the *identity* transformations:

(i) *The null matrix:* $\mathbf{T} = \begin{pmatrix} 0 & 0 \\ 0 & 0 \end{pmatrix}$

projects all points to the origin (0, 0) of the space and in so doing destroys all information (and obviously preserves neither distances nor scalar products),

e.g. $\underset{\mathbf{T}}{\begin{pmatrix} 0 & 0 \\ 0 & 0 \end{pmatrix}} \underset{\mathbf{P}}{\begin{pmatrix} 2 \\ 1 \end{pmatrix}} \qquad = \begin{pmatrix} 0 \\ 0 \end{pmatrix}$
$= \mathbf{O} \text{ (origin)}$

(ii) *The unit matrix:* $\mathbf{T} = \begin{pmatrix} 1 & 0 \\ 0 & 1 \end{pmatrix}$

simply preserves the exact location of all points and is therefore an identity mapping. It preserves all information (including distances and scalar products),

e.g. $\underset{\mathbf{T}}{\begin{pmatrix} 1 & 0 \\ 0 & 1 \end{pmatrix}} \underset{\mathbf{P}}{\begin{pmatrix} 2 \\ 1 \end{pmatrix}} \qquad = \begin{pmatrix} 2 \\ 1 \end{pmatrix}$
$= \mathbf{P}'$

A7.1.4 Diagonal transformations (reflection and rescaling)

A particularly important set of geometric transformations involve diagonal transformation matrices (whose off-diagonal elements are zero). Clearly, the unit matrix is one example; other relevant instances are:

   reflection;
   uniform rescaling;
   differential rescaling (weighting of axes);

(i) *Reflection*

Reflection of points in 2-space occurs when the sign $(+, -)$ of *one* of the co-ordinate axes is changed. The effect is to 'flip' the configuration symmetrically. This is achieved by a diagonal transformation matrix, *one* of whose elements is -1.

| *Reflection in y-axis* | *Reflection in x-axis* |

$\mathbf{T} = \begin{pmatrix} -1 & 0 \\ 0 & 1 \end{pmatrix}$ $\qquad\qquad\qquad$ $\mathbf{T} = \begin{pmatrix} 1 & 0 \\ 0 & -1 \end{pmatrix}$

Note Reflection preserves distances and scalar products identically.

Example $P = (2, 1)$ and $Q = (4, 2)$ (see Figure A7.3)

   (a) *Reflection in y-axis*

$\underset{\mathbf{T}}{\begin{pmatrix} -1 & 0 \\ 0 & 1 \end{pmatrix}} \underset{\mathbf{P}}{\begin{pmatrix} 2 \\ 1 \end{pmatrix}} = \underset{\mathbf{P}'}{\begin{pmatrix} -2 \\ 1 \end{pmatrix}} \text{ and } \underset{\mathbf{T}}{\begin{pmatrix} -1 & 0 \\ 0 & 1 \end{pmatrix}} \underset{\mathbf{Q}}{\begin{pmatrix} 4 \\ 2 \end{pmatrix}} = \underset{\mathbf{Q}'}{\begin{pmatrix} -4 \\ 2 \end{pmatrix}}$

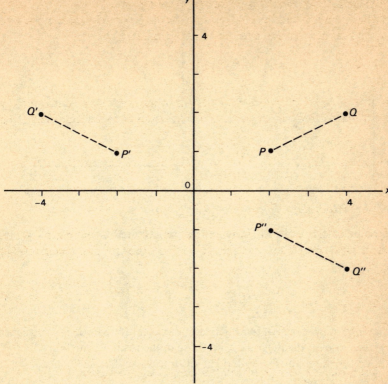

Figure A7.3 *Reflection on x-axis*

(b) *Reflection in x-axis*

$$\begin{pmatrix} 1 & 0 \\ 0 & -1 \end{pmatrix}\begin{pmatrix} 2 \\ 1 \end{pmatrix} = \begin{pmatrix} 2 \\ -1 \end{pmatrix} \text{ and } \begin{pmatrix} 1 & 0 \\ 0 & -1 \end{pmatrix}\begin{pmatrix} 4 \\ 2 \end{pmatrix} = \begin{pmatrix} 4 \\ -2 \end{pmatrix}$$
$$\quad \text{T} \quad \text{P} \ = \ \text{P''} \qquad\qquad \text{T} \quad \text{Q} \ = \ \text{Q''}$$

(ii) *Uniform weighting (rescaling)*
A diagonal transformation matrix *with identical diagonal elements a,*

$$\text{T} = \begin{pmatrix} a & 0 \\ 0 & a \end{pmatrix}$$

produces a uniform stretching ($a > 1$) or shrinking ($0 < a < 1$) of point locations, with each axis weighted by a factor of a. The effect is to rescale a configuration by a factor of a.

Note Uniform weighting/rescaling preserves *relative* distances and scalar products, up to a proportional rescaling by a, i.e. a ratio scale.

Example $\text{T} = \begin{pmatrix} 2 & 0 \\ 0 & 2 \end{pmatrix}$

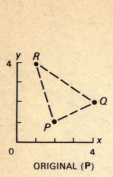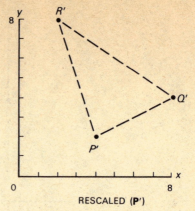

ORIGINAL (P)   RESCALED (P')

Figure A7.4   *Rescaling with uniform weighting*

and the point co-ordinates for P, Q and R are gathered into a matrix **P**:

$$
\begin{array}{cc}
 & P\ Q\ R \qquad\quad P'\ Q'\ R' \\
\begin{pmatrix} 2 & 0 \\ 0 & 2 \end{pmatrix} \begin{pmatrix} 2 & 4 & 1 \\ 1 & 2 & 4 \end{pmatrix} = \begin{pmatrix} 4 & 8 & 2 \\ 2 & 4 & 8 \end{pmatrix} \\
\quad\ \mathbf{T} \qquad\quad \mathbf{P} \qquad = \qquad \mathbf{P'}
\end{array}
$$

In the distance matrix below, the Euclidean distances in the *original* configuration (**P**) are given beneath the diagonal, and those in the rescaled configuration (**P'**) are given above the diagonal. Note that the distances are doubled, since $a = 2$.

$$
\mathbf{D} = \begin{pmatrix}
 & P & Q & R \\
P & 0 & 4.47 & 6.32 \\
Q & 2.24 & 0 & 7.21 \\
R & 3.16 & 3.61 & 0
\end{pmatrix}
$$

(iii)   *Differential weighting (rescaling)*
A diagonal transformation matrix whose diagonal values are *not* equal

$$
\mathbf{T} = \begin{pmatrix} a & 0 \\ 0 & b \end{pmatrix} \quad \text{with} \quad a \neq b
$$

produces a differential elongation or compression of the X and Y axes, with the X-axis weighted by a and the Y-axis by b. The effect is to distort the configuration along the directions of the axes, increasing the co-ordinate values if the weight is greater than 1, and decreasing them if it is less than 1. In the process of differential rescaling the original distances are changed, often dramatically so if the weights are very different.

Note   Differential weighting does *not* preserve even relative distances or scalar products. It is not therefore a similarity transformation (see below).

$$
\begin{array}{cc}
 & P\ \ Q\ \ R \qquad\quad P''\ Q''\ R'' \\
\begin{pmatrix} \frac{1}{2} & 0 \\ 0 & 3 \end{pmatrix} \begin{pmatrix} 2 & 4 & 1 \\ 1 & 2 & 4 \end{pmatrix} = \begin{pmatrix} 1 & 2 & \frac{1}{2} \\ 3 & 6 & 12 \end{pmatrix} \\
\quad \mathbf{T} \qquad\quad \mathbf{P} \qquad = \qquad \mathbf{P''}
\end{array}
$$

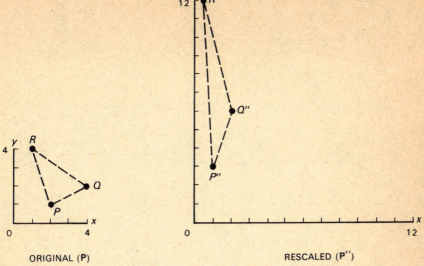

ORIGINAL (P)   RESCALED (P'')

Figure A7.5   *Rescaling with differential weighting*

$$\mathbf{D} = \begin{array}{c} \\ P \\ Q \\ R \end{array} \begin{array}{ccc} P & Q & R \\ \left(\begin{array}{ccc} 0 & 3.16 & 9.01 \\ 2.24 & 0 & 6.19 \\ 3.16 & 3.61 & 0 \end{array} \right) \end{array}$$

(original distances below diagonal, rescaled above diagonal)

A7.1.5   (Orthogonal) rotations

Very often the co-ordinate axes of a set of points are arbitrary, especially for Euclidean distance calculations, and another set of co-ordinate axes (such as principal components) may be preferable or have more desirable properties. The move from one set of co-ordinate axes to another is termed a *rotation of axes* about the origin of the space. In the 2-dimensional case, this move is described in terms of the *direction* and the *extent* of the change. (Here we assume a rigid or *orthogonal* rotation, keeping the axes at right angles.) The axes may be moved clockwise, or in an anticlockwise direction, and the extent of the rotation is given by the angle θ through which the axes move.

(i)   *Permutation of axes.*

A rigid rotation anticlockwise through $90°$ is effected by a rotation matrix of the form $\mathbf{T}_a$. (where the subscript a denotes an anticlockwise rotation)

$$\mathbf{T}_a = \begin{pmatrix} 0 & 1 \\ -1 & 0 \end{pmatrix}$$

This has the effect of permutating the x into the y axis, or moving the co-ordinate system through a right angle. The rotation matrix, $\mathbf{T}_c$, (where the subscript c denotes a clockwise rotation)

$$\mathbf{T}_c = \begin{pmatrix} 0 & -1 \\ 1 & 0 \end{pmatrix}$$

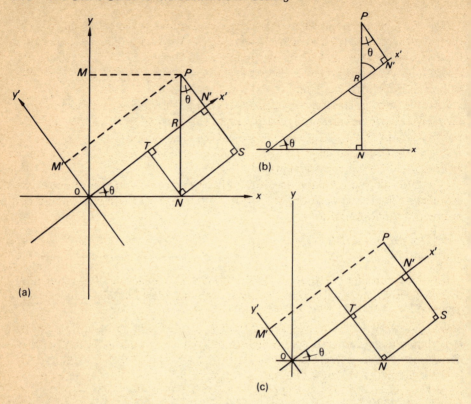

Figure A7.6   *The geometry of rotation of axes*

has the opposite effect, rotating the axes in a clockwise direction through 90°.

(ii)   *Rotation through an angle*
For any angle other than 90°, the anticlockwise rotation matrix $\mathbf{R}_a$, taking the original x, y axes into the new x', y' axes has the form:

$$\mathbf{R}_a = \begin{pmatrix} \cos\theta & \sin\theta \\ -\sin\theta & \cos\theta \end{pmatrix}$$

where θ is the positive angle through which the original reference axes are rotated. (Readers who wish to ignore the trigonometry of this rotation matrix should skip the following section and move to the example, simply noting that if $\theta = 90°$, then $\cos\theta = 0$ and $\sin\theta = 1$, reducing $\mathbf{R}_a$ to the simple form $\mathbf{T}_a$ noted earlier.)

How is the rotation matrix derived? Although not obvious, the procedure is quite simple. You are advised to follow the steps in Figure A7.6 (a)–(c).

First we do some elementary geometry.

(1)   We identify a point P with reference to the old axes (x, y) and the new axes $(x'$ $y')$ which have been rotated through an angle of θ degrees.

(2)   The co-ordinates of P on x and y are N and M respectively, its co-ordinates on x' and y' are respectively N' and M'.

(3)   It is useful to convince yourself at this stage that the length $OM =$ the length PN and, similarly $MP = ON$, $OM' = N'P$ and $ON' = M'P$.

(4)   We label the point where PN crosses x' as R for convenience, and consider the two triangles ORN and RPN' (see inset b). We know that the angles of a triangle sum to 180°, and therefore the angle ORN must be $180° - (90° + \theta°)$ $= 90° - \theta°$. The angle PRN' is also $90° - \theta°$, thus the angle RPN' must be $\theta°$. We will make use of this fact later.

(5)   We now construct a rectangle $TN'SN$, redrawn for clarity in inset c. Again the reader should be convinced that the length ON' is equal to $OT + TN'$ and ON' *is also equal to* $OT + NS$ (since $TN' = NS$ by construction).

(6)   Also, OM' is equal to $PS - SN'$ and OM' is also equal to $PS - TN$ (since again $SN' = TN$ by construction).

(7)   Returning to the large diagram (inset a) we are now in a position to give expressions for the new co-ordinates of P (N' and M') in terms of the old co-ordinates (N and M), thus

$$ON' = OT + NS$$
and $$OM' = PS - TN$$

(8)   Now consider the triangles OTN and PNS. The reader should be convinced that these right-angled triangles are similar, having identical angles (90°, θ and $90° - \theta°$).

(9)   We know that in a right-angled triangle the cosine of an angle θ is given by the ratio of the side adjacent to the angle to the hypotenuse. Thus, in triangle OTN the cosine of θ (cos θ) is OT/ON, and in PNS, cos $\theta = PS/PN$.

(10)   Similarly, we know that the sine of an angle (sin θ) is given by the ratio of the side opposite the angle to the hypotenuse. Thus, in PNS the sine of θ (sin θ) is NS/PN, while in OTN, sin $\theta = TN/ON$.

(11)   We are now in a position to do some algebra. Consider the fact that

$$ON' = OT + NS \tag{1}$$

We seek an expression for ON' in terms of O, N, M, P and the angle θ (i.e. an expression which will relate the new axes to the old).
We know that

$$\cos \theta = \frac{OT}{ON}.$$

Cross-multiplying gives

$$OT = ON \cos \theta. \tag{2}$$

Similarly we know

$$\sin \theta = \frac{NS}{PN}.$$

Again, cross-multiplying gives

$$NS = PN \sin \theta. \tag{3}$$

Thus, substituting (2) and (3) back in (1), we obtain

$$ON' = ON \cos \theta + PN \sin \theta. \tag{4}$$

We may treat the expression

$$OM' = PS - TN$$

in precisely the same way.

$$\cos \theta = \frac{PS}{PN}$$
$$PS = PN \cos \theta$$
$$\sin \theta = \frac{TN}{ON}$$
$$TN = ON \sin \theta$$

Therefore
$$OM' = PN \cos \theta - ON \sin \theta \qquad (5)$$

Thus in (4) and (5) we have derived the desired result: an expression for the new co-ordinates in terms of the old co-ordinates and the angle of rotation.

Now, letting x stand for ON, y for OM, x' for ON' and y' for OM' we may write

$$\left. \begin{array}{l} x' = x \cos \theta + y \sin \theta \\ y' = -x \sin \theta + y \cos \theta \end{array} \right\}$$

which in matrix form is

$$\begin{pmatrix} x' \\ y' \end{pmatrix} = \begin{pmatrix} \cos \theta & \sin \theta \\ -\sin \theta & \cos \theta \end{pmatrix} \begin{pmatrix} x \\ y \end{pmatrix}$$

i.e.

$$\mathbf{x'} = \mathbf{R}_a \mathbf{x}$$

which is the desired result.

An example will illustrate the procedure. In Figure A7.6a, the angle of rotation ,θ, is almost $37°*$. In this instance, the position of P with respect to the new axes (X', Y') is obtained as follows:

$$\begin{pmatrix} x' \\ y' \end{pmatrix} = \begin{pmatrix} \cos 37° & \sin 37° \\ -\sin 37° & \cos 37° \end{pmatrix} \begin{pmatrix} 4.0 \\ 5.2 \end{pmatrix} = \begin{pmatrix} 6.32 \\ 1.75 \end{pmatrix}$$

If we wish to rotate a whole configuration of P points in 2 dimensions, the same rotation matrix is applied to each point. (In more than 2 dimension the rotation procedure is slightly more complicated, but consists of taking all pairs of dimensions and proceeding in the above manner.)

A7.1.6 Transformations: a summary
In discussing the various geometrical transformations employed in MDS, the following are particularly significant and basic:

The *identity or unit* transformation, which leaves a configuration and its absolute and relative distances and scalar products unchanged;

The *(central) dilation or rescaling* transformation, which multiplies each axis by a

---

*In fact it is $36°52'$; the triangle ORN is a 3:4:5 triangle, yielding a cosine of 4/5 and a sine of 3/5.

given scalar value. It preserves relative (but not absolute) distances and scalar products.

The *differential dilation or rescaling* transformation, which multiplies each axis by a different scalar value. It preserves neither relative nor absolute distances (nor scalar products).

The *translation* transformation which removes the origin of the space. It preserves relative and absolute (Euclidean) distances, but changes scalar products.

The *orthogonal rotation* transformation, which preserves Euclidean distances and scalar products identically.

An *extended similarity* transformation (often called a similarity transformation) is a composite transformation, which preserves all relative distances. It may comprise a reflection, a translation of origin, a central dilation and an orthogonal rotation. Taken together these are permissible operations on a distance model configuration.

APPENDIX 7.2   NOTES ON THE ESTIMATION PROCEDURE IN INDSCAL

Full details of the alternating least squares procedure for estimating the parameters of the INDSCAL model are contained in Carroll and Chang (1970), Carroll and Wish (1973) and in the MDS(X) documentation of INDSCAL-S.

What follows here is a brief introduction to the basic method of analysis.

(i)   The basic model assumes that the subject's data dissimilarities are a linear function of the distances of the solution

$$\delta_{jk}^{(i)} = L(d_{jk}^{(i)})$$

where the distances refer to the *i*th individual's (private) space, i.e.

$$d_{jk}^{(i)} = \sqrt{\sum_a (y_{ja} - y_{ka})^2}$$

The first step, as in other metric models, is to *convert the subject's data ('relative distances') into estimates of distances* by calculating an additive constant, as in classic scaling (see 5.2.3.2), which will make the data satisfy the triangle inequality.

(ii)   *The data 'distances' are then converted into estimated scalar products,* as in classic metric scaling (see Appendix A5.2), with their origin at the centroid of the points. At this stage, each subject's data are normalised to have equal influence. The relationship between the estimated scalar products (b_{jk}) and the private space co-ordinates is simply:

$$b_{jk}^{(i)} = \sum_a y_{ja}^{(i)} y_{ka}^{(i)} \tag{1}$$

(iii)   The INDSCAL model, stated in its distance form is:

$$\delta_{jk}^{(i)} = \sqrt{\sum_a w_a^{(i)} \left(x_{ja} - x_{ka} \right)^2}$$

and the relationship between the group space co-ordinates (x_{ja}) and the private space co-ordinates (y_{ja}) is:

$$y_{ja}^{(i)} = \sqrt{w_a^{(i)}} x_{ja} \tag{2}$$

Substituting (2) into (1) gives

$$b_{jk}^{(i)} = \sum_a w_a^{(i)} x_{ja} x_{ka} \tag{3}$$

This is the three-way scalar products formulation of the INDSCAL *model.* For notational simplicity, it helps to rewrite (3), putting subject references (i) as subscripts:

$$b_{ijk} = \sum_a w_{ia} x_{ja} x_{ka} \tag{3a}$$

(iv)   The estimation of the subject weights (w_{ia}) and group space co-ordinates (x_{ja}) in INDSCAL is performed by a variant of the three-way canonical decomposition model (see 7.2.2), which ensures that the second and third ways $(x_{ja}$ and $x_{ka})$ are in fact identical.

(v)   The INDSCAL model has been shown by Schönemann (1972) to have an exact algebraic solution—for perfect data. In the case of errorful data, an iterative process (which may use Schönemann's method to provide an initial configuration) is employed, using an alternating procedure. It consists of finding a preliminary estimate for the two stimulus weights $(x_{ja}$ and $x_{ka})$, fixing them, and then estimating (by least squares) the subject weights w_{ia}. Then the x_{ja} are estimated, with the w_{ia} and x_{ka} fixed, and so on.

When a satisfactory approximation to the data is obtained, the process terminates, ways 2 and 3 are set equal, a final estimate of way 1 (subject weights) is made and the weights are then appropriately normalised before being output.

A caution
All variants of alternating least squares estimation procedures are susceptible to a greater or lesser extent to local minimum solutions. In any event, users should be prepared for this eventuality: often ten runs with different starting configurations are necessary before one can be virtually certain that one has an optimal solution. In any event, it would be foolhardy to rely on less than three. In the repeated runs, the group space configurations will probably be very similar *except for slight differences in orientation.* Since subject weights refer directly to a particular orientation and will often change considerably under relatively small rotations of the dimensions, particular attention should therefore be paid to how the group space dimensions change and to the individual and overall goodness of fit measures.

Developments in Multidimensional Scaling

8.1 Introduction

Prediction is a hazardous business. Nonetheless, a number of major trends in MDS developments over the past few years merit attention and seem to point towards a convergence of interest between spatial (MDS) and non-spatial (clustering) models.

The MDS(X) package both responds to, and in part helps create such developments; it would be gratifying to think that there is a self-fulfilling prophecy involved. But three major areas of development seem evident:

developments in INDSCAL and three-way scaling;
a shift from exploratory towards confirmatory scaling;
developments in discrete (clustering) and hybrid MDS/clustering procedures.

We will consider each of these in turn.

8.2 Developments in INDSCAL and its Estimation*

The basic INDSCAL model was described earlier in 7.2. The simplest version involves two distinct configuration matrices (sometimes called **X** left and **X** right to distinguish them), although they are set equal in the two-mode INDSCAL-S program. In slightly different notation than that used before, this basic model can be defined as follows:

General distance

$$d_{jk}^{(i)} = \sqrt{\sum_{aa'} (x_{ja} - x_{ka})c_{aa'}^{(i)}(x_{ja'} - x_{ka'})} \qquad \text{(1a)}$$

General scalar products

$$b_{jk}^{(i)} = \sum_{aa'} x_{ja}c_{aa'}^{(i)}x_{ka'} \qquad \text{(1b)}$$

*This section is based upon Coxon and Jones 1980, p. 48 et seq.

... or in matrix notation

$$\mathbf{B} = \mathbf{X}\mathbf{C}_i\mathbf{X}' \qquad (1c)$$

When the simple form of the INDSCAL model is stated in matrix form, the transformation matrix $\mathbf{C}_i$ is a diagonal matrix whose entries are the individual dimension-weights. However, if other models of individual differences are desired, there is no reason why $\mathbf{C}_i$ has to take this form. One possibility is to define $\mathbf{C}_i$ as an idiosyncratic rotation of the $\mathbf{X}$ matrix, combined with a weighting transformation as before. This is shown in equation (2) and defines the Carroll-Chang ISDIOSCAL model referred to in the last chapter. On the other hand, if a

$$\mathbf{C}_i = \mathbf{T}_i\mathbf{W}_i\mathbf{T}_i' \text{ with } \mathbf{T}_i \text{ orthogonal (rotation matrix)} \qquad (2)$$
$$\text{and } \mathbf{W}_i \text{ diagonal (individual weights)}$$

model without any individual differences is required, then when $\mathbf{C}_i$ is the identity matrix, and we have a scalar products version of the simple metric MDS model.

However, providing exotic models of individual differences must eventually give way to finding methods for parameter estimation which give reliable results and are not subject to local minimum problems. Bloxom (1974) has suggested that Jöreskog's general method for the analysis of covariance structures (ACOVSM) can be used to estimate the group space and the subject weights from the data in scalar products form. According to Bloxom, this use of ACOVSM should be computationally more efficient when there are a large number of individuals. A further point is that the estimation procedure is a very well-understood one.

Probably the most important development was Schönemann's (1972), referred to in A7.2 above. It is well known that canonical decomposition as a method of solution for the INDSCAL model does not involve a well-defined minimising criterion, at least not in terms of the weighted distance model. (It does minimise *strain*, defined in terms of the scalar products computed from the data; see Carroll and Wish, 1973, p. 89.) Schönemann presents an analytic solution to the subjective metrics model, based upon earlier work by Meredith, who was seeking a criterion for rotational invariance in factor analysis. While Schönemann's solution holds only for perfect data, a variant of it has been included in almost all subsequent iterative procedures for estimating INDSCAL parameters, including the MDS(X) version of INDSCAL-S. Briefly, Schönemann argues that an optimal solution can be obtained for equation (2) (and hence (1c)) by obtaining an estimate for the overall orientation of the master space $\mathbf{T}$, by choosing the average subject and an arbitrary second subject (see Carroll and Wish 1974p, p. 441). This holds because two symmetric Gramian matrices are simultaneously diagonisable by a single transformation matrix. Even for errorful data, Schönemann's analytic solution has been shown to improve estimates considerably.

The North Carolina-Leiden team (Young, Takane and de Leeuw) have produced a fairly robust and efficient procedure, based upon ALS (Alternating Least Squares) and minimising a criterion ('SSTRESS'), defined in terms of the proportion of variance in the model due to incorrectly ordered pairs of distances, relative to the ordering of the dissimilarities (Takane et al. 1977, p. 17). In so doing, this ALS procedure computes Schönemann's solution as an initial estimate, obtains optimally scaled (least squares) of disparities in one phase, and calculates new conditional least squares estimates of the weights (conditional on the configuration co-ordinates X, and the fitted disparities $\hat{D}$) and the co-ordinates X (conditional on the weights and disparities). This second phase is implemented by a fairly efficient, but rather slow, modification by Gill and Murray of the Newton-Raphson procedure. It should be noted that this ALS procedure moves directly to a non-metric solution.

8.2.1 Ramsay's MULTISCALE models

Parallel with the important work in the ALS tradition, impressive results have been achieved by Ramsay (1977) in his work on obtaining maximum likelihood estimations in the conventional (simple) and weighted distance models. Ramsay notes that the hidden assumption in the least-squares fitting criterion used in most MDS is that (possibly derived) scalar products are independently, and normally, distributed about their 'true' values, whilst scalar products are usually based on the squares of the original dissimilarities data, and therefore inevitably have higher standard error. He also argues against the unthinking use of non-metric fitting techniques on the grounds that they rapidly consume degrees of freedom. He suggests that the lognormal distribution is a good candidate for the distribution of a dissimilarity around its (errorless) value, that it is possible to estimate the configuration X independently of the estimation of the standard error, and that a power transformation is probably a better transformation (psychologically considered) than either a linear or a monotone one.

Ramsay's work is instructive, since it shows three variations on the basic MDS model, and a simultaneous concern with issues in statistical inference. Since Ramsay's substantive interests are in the psychophysics of human judgment, his models assume that the proximities data between pairs of points have been obtained by asking people to make direct judgments of the amount of dissimilarity between pairs of stimuli, usually in a conventional rating-scale format. It is important to note that the models are metric, and that they assume a lognormal distribution for the errors of estimation, that is, they assume that the standard error of estimate for the dissimilarity between any pair of points will be proportional to the errorless (population) value of that dissimilarity. The MULTISCALE package* estimates four MDS models, of which the first (M1) is the basic metric MDS model. The remaining models (M2 through M4) assume that each subject's dissimilarities data have a power law relationship to the distances in the solution configuration (for a review of the power law in psychophysics, see Stevens, 1966). Model M2 is:

$$d_{jk}^{(i)} = v^{(i)} \left(\sum_{a=1}^{r} (x_{ja} - x_{ka})^2 \right)^{p^{(i)}/2} \tag{3}$$

where the exponents $p^{(i)}$ are estimated for each subject, and thus allow each subject to have a different power law relationship to the common distances in the solution space. Ramsay notes that the value of the exponent can be a useful clue in

*International Educational Services,
 1525 East 53rd St., Room 829,
 Chicago, Illinois 60615, USA.
 International Educational Services distribute Ramsay's MULTISCALE set of four computer programs.

exploratory data analysis: specifically, a subject whose exponent value is less than about 0.3 may be one whose data are markedly inconsistent with the MDS model. Model M2 also estimates a scale parameter, $v^{(i)}$, for each subject. This does not usually have any interest, as it is merely a ratio scale transformation (like the translation from miles to kilometres) to allow for the possibility that some subjects may use larger or smaller average ratings than others.

It is an indication of the general shift from exploration to confirmation that these models are estimated with a great deal of concern for the goodness of fit between data and solution in relation to both the number of parameters estimated by the model, and the *shape* of the distribution of residuals. If $d_{jk}^{(i)}$ is the distance between j and k in the proximities data, and $d_{jk}^{(i)}*$ is the corresponding fitted value, and $EDF^{(i)}$ is the number of degrees of freedom for error, when the model is fitted to the i-th subject, then the unbiased estimate of the standard error of prediction associated with that subject is:

$$\sigma^{(i)} = \sqrt{\left\{ \frac{\sum\limits_{jk} (\log d_{jk}^{(i)} - \log d_{jk}^{*(i)})^2}{EDF^{(i)}} \right\}} \qquad (4)$$

An estimate for the overall standard error of prediction can be obtained by taking the expectation of $\sigma^{(i)}$ over all subjects.

One of the attractive features of Ramsay's approach is that it allows the application of statistical tests between pairs of models where one is a special case of the other. In itself, this is nothing new, since, as we have seen for PREFMAP and PINDIS, any nested pair of metric MDS models which are estimated with a least squares loss function can be subjected to an F-test for the statistical significance of the difference in explained variance. What is novel in Ramsay's approach is his theoretical rationale for assuming that errors have a lognormal distribution, and his inclusion of diagnostic plotting options in order to check up on this assumption. Since model M2 with a smaller number of dimensions is a special case of the same model with a larger number of dimensions, it is possible to make a statistical test as to which number of dimensions should be preferred. The statistical tests are made with chi-square, derived from a comparison of log-likelihood values associated with each of the two solutions to be compared. Examples are given in Ramsay (1977) and Coxon and Jones (1980, pp. 54-9).

It is convenient to review Ramsay's model M4 before discussing his version of individual differences scaling (model M3). An interesting feature of model M4 is that its expected values are exactly the same as for model M2. The difference is that the standard errors of estimation are expected to vary systematically, depending upon which stimuli are being compared. Suppose, for example, that the stimuli were candidates for political office, some being well known, perhaps running for re-election, and others being unknown outsiders. Opinion might be much more homogeneous about the properties of any well-known candidate than about those of any outsider. The discrepancy between data and fitted values should be smaller for pairs of stimuli about whose characteristics people have homogeneous opinions than for pairs of stimuli involving at least one lesser-known stimulus. In the spirit of Thurstone's estimation of discriminal dispersions, in a different context, Ramsay suggests that the standard error of prediction between data and fitted values for any given pair of stimuli (over all replications) be taken as:

$$\sigma_{jk} = \sqrt{\sigma\left(\frac{\alpha_j^2 + \alpha_k^2}{2}\right)} \tag{5}$$

where σ is the common standard error, and α_j and α_k are the standard error weights for stimuli j and k. The standard error weights are constrained to have a mean of unity, so that stimuli whose weights are larger than 1.0 reveal non-homogenous responses by the subjects, with homogeneous responses indicated by weights less than unity.

Model M3 is also a more general case of M2, but has no nesting relation with M4. It is also a variation of the weighted individual differences (INDSCAL) model, as is apparent from equation below:

$$d_{jk}^{(i)} = v^{(i)}\left(\sum_{a=1} w_a^{(i)}(x_{ja} - x_{ka})^2\right)^{p^{(i)}/2} \tag{6}$$

where the symbols are as explained in equation (3), with $w_a^{(i)}$ being the weight which individual i's give to dimension a. As in all scaling models, it is necessary to place some constraints on possible values of the parameters. In the equation for model M3, if all that is required is to maintain the equality between left and right hand sides, then any multiplication of the weights $w_a^{(i)}$ for any subject can be offset by a counterbalancing change in the regression coefficient $v^{(i)}$ for that subject. Similarly, the co-ordinates of the points in the common configuration **X** can be made arbitrarily large or small, as long as counterbalancing changes are made in the regression coefficients or the subject weights. To avoid these indeterminacies, constraints are placed on the parameter estimates.

As a result of the constraint on the sum of weights, the subject weights estimated in model M3 are slightly different from those estimated in the original Carroll-Chang INDSCAL model, which has neither a regression coefficient nor an exponent for each subject, and which normalises the weights for any subject so that their sum of squares is equal to the proportion of explained variance for that subject. This is important, since researchers who are concerned with individual differences usually wish to take the subject weights as an index of differences in judgment, and to use them in further statistical analysis. For this and other reasons referred to below, the strong intuition that it is hazardous to indulge in detailed analysis of subject weights seems well founded. MacCallum (1977), as we saw in 7.2.1.2, would go further and forbid the use of the general linear model, in any form, as an aid to interpreting subject weights. Users have been warned!

8.3 Confirmatory Scaling

In addition to their work on individual differences scaling, Borg and Lingoes (1980) have devised a practical method for imposing theoretical constraints on a scaling solution in a program known as CMDA (Confirmatory MultiDimensional Analysis). Side-constraints are imposed on the distances of the solution configuration **X**. These constraints must be derived from the researcher's substantive theory, and are of course additional to the constraints imposed on the solution distances by the data matrix of proximities. Such extra constraints are highly desirable if the data are patterned in such way that qualitatively different solutions can have similar overall stress values. This unfortunate situation is to be expected when scaling small numbers of points (less than 12 as a rough rule of thumb), but can also arise with larger problems. (Borg and Lingoes provide an example in the scaling of 16 points.) Constraints on the solution distances are expressed in one or more matrices of 'pseudo-data', and relations among elements within these pseudo-data matrices express the restrictions to be imposed on corresponding solution distances. Overall stress will obviously be higher for a constrained than for an unconstrained solution, and particular points will usually—though not always—be worse fit after the theoretical constraints have been applied. Such a confirmatory approach can only be applied in areas where strong theories are available to guide the hypotheses incorporated into the constraints. Borg and Lingoes give examples in which the hypothesis is 'structural' or derived from facet theory. This hypothesis is that usually a simplex or circumplex pattern underlies the data.

In the case of three- (and higher) way scaling, similarly constrained models have been developed by Carroll and his collaborators, known under the generic title of CANDELINC—CANonical DEcomposition and LINear Constraints on the parameters—(Carroll and Green 1976). Whether the desire is to set constraints only on the stimuli (or other single way of the model) or upon the full set, CANDELINC allows the user to test out hypotheses or estimate the effects of specified designs on INDSCAL and allied analyses. (It is particularly apposite to note that linear constraints turn out to be singularly inappropriate for constraining subject vectors, as in the MDPREF and INDSCAL models.) The less-developed social sciences may have to improve the quality of their theories in order to reap the benefits of this sort of confirmatory scaling methodology.

8.4   Developments in discrete (non-spatial) models

Although the main focus in this book has been upon spatial models for scaling similarity data, other non-spatial models also exist and have developed in parallel with MDS. Clustering and classification methods (Cormack 1971, Wishart 1978, Everitt 1974) are as extensive as scaling methods, and in Chapter 4 we encountered both hierarchical clustering schemes and graphs as representations of similarity data.

In recent years non-spatial 'discrete' representations of similarity data have been extended by two major developments—additive trees and additive (overlapping) clustering procedures.

8.4.1   Additive trees

A *tree* is a graph consisting of a set of p points (or nodes) connected by (p-1) links (arcs or edges), which has no cycles (Flament 1963, pp. 23–4). Consequently any two points in the tree are connected by only one path. An *additive tree* is a tree where the arcs have non-negative, real-valued lengths in which the distance between two points is the length of the path (i.e. the sum of the arc lengths) joining them.* Buneman (1971) showed that an additive tree must satisfy a so-called four-point condition which is more exacting than the triangle inequality.

Given any four distinct points (i, j, k, l) and the six dissimilarities ('distances') between them, the necessary and sufficient condition for an additive tree representation is that:

$$(d_{ij} + d_{kl}) \leqslant \max \{(d_{ik} + d_{jl}), (d_{jk} + d_{il})\}.$$

Notice that if k is put equal to l, then the condition reduces to the ordinary triangle inequality; it can also be shown (cf. Buneman 1971, p. 392) that an ultrametric satisfies the four-point condition, so hierarchical clustering schemes are a special case of additive trees. A practical algorithm to produce an additive tree (especially from errorful, genuine data) took some time to produce, but currently workable variants (Cunningham 1978; Sattath and Tversky 1977) all follow the same basic procedure, which is fully discussed in Cunningham (1978).

(i) Determine the linear constraints implied by the four point condition.
(ii) Find a set of distances satisfying those constraints which are as close as possible to the original data dissimilarities, in the least-squares sense that: $\sum (d_{ij} - \delta_{ij})^2$ is a minimum.
(iii) Construct the additive tree which generates the distances.

Several comments are in order. The implication is that the data are at least interval level, and the algorithm is therefore a metric procedure.

However, if the data are ordinal and the researcher wishes to employ a quasi non-metric variant, then the data (dis)similarities may be pre-processed using MVNDS (6.1.7 q. v.) to produce distances which are guaranteed to satisfy at least the triangle inequality, and these may then be input to the additive tree program. Incidentally, this provides an excellent example of the general utility of the MVNDS program.

Secondly, when a tree or graph is drawn on a sheet of paper the actual positioning of the points is entirely arbitrary from the standpoint of graph theory, so a particular additive tree can be drawn in a number of apparently different ways. But, as in the case of other non-spatial models such as HCS and MVNDS, it is often helpful and illuminating to begin by scaling the data in two dimensions, using SSA, in order to provide a configuration of point locations into which the additive tree representation can then be embedded (see Figure 8.1, reproduced from Shepard 1980, p. 395 for an example).

Thirdly, the algorithms for producing an additive tree are sometimes liable to give rise to degenerate or trivial representations of the data. Whether this is a temporary shortcoming of the computational methods or whether it arises from the specific properties of the model is not entirely clear. In any event, the Cunningham-Shepard algorithm guarantees that only violations of the smallest magnitudes will occur, and the studies of Sattath and Cunningham have shown that, compared to both HCS and SSA, additive tree analysis does remarkably well in accounting for the original variation in the data in an efficient manner.

*The most detailed treatment of additive tree representation of similarity is given in Sattath and Tversky (1977) and Cunningham (1978), who build upon earlier work by Buneman (1971), Carroll and Chang (1973) and others. An excellent overview is given in Shepard (1980).

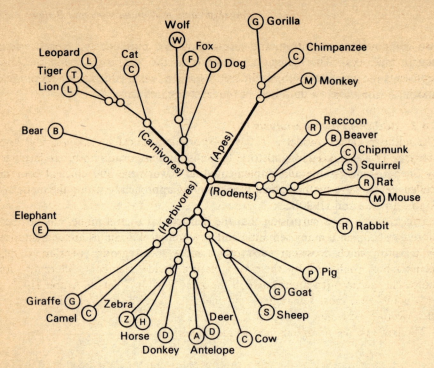

Figure 8.1 *Additive tree of Henley's data on the conceptual similarities between 30 species of animal, embedded in a 2-dimensional space*

Applications of additive tree analysis have so far been reported as illustrative examples in the methodological literature cited above. It seems clear, however, that it is especially suited to the representation of semantic similarity structures, and a number of interesting applications are reported by Sattath and Tversky, (including judged similarity of Henley's animal terms (reproduced in Figure 8.1), Swedish lower-case letters, and occupational titles) as well as by Cunningham (1978), Shepard (1980), and Carroll (1976).

A particularly interesting non-psychological application is the one originally envisaged by Buneman (1971), the filiation of manuscripts. Given a number of copies of a particular text, is it possible to infer the 'family tree' which generated the variants? If so, it will be possible (in principle) to decide, by the assumption of common errors, which of them is the original and how the later copies relate to it. Since an additive tree may, and usually will, represent some objects at levels other than end-nodes (unlike HCS where they must all be located as end-nodes), this property makes it an obvious candidate for such investigations. The practicalities of the filiation procedure and examples of its use are discussed in the articles following Buneman's in Hodson et al. (1971).

Cunningham (1978) has extended additive tree analysis to deal with square asymmetric data, representing the asymmetric by a bi-directional graph where d_{ij} and d_{ji} have distinct (and usually differently valued) arc-lengths. (This has turned out to be particularly useful for representing asymmetric data of the sort discussed under 5.1.1.1.) Carroll and Pruzansky (1975, p. 3) have also generalised additive tree analysis to the 3-way case, showing that a path matrix can be decomposed into two components—an ultrametric distance matrix, plus one (or more) residual matrix (matrices). The former corresponds to the 'group tree' and the latter represents the individual subject differences which, when added to the group path matrix, fit the subject's data best in the least squares sense.

8.4.2 Additive cluster analysis

The additive cluster model is a type of non-hierarchical *overlapping* clustering analysis. This also has a long history, but in its present form it has been developed by Shepard and Arabie and implemented as a workable and efficient program (called MAPCLUS—a mathematical programming approach to fitting the model) by Arabie and Carroll (1980).

It is in many ways surprising that the demand for such a model has not been more insistent, since many semantic and psychological domains are characterised by features which cross-cut each other; most social groups overlap in their membership; coalitions by their very nature divide members differently; and biological species often intermingle in the same territory. Yet, as Arabie (1977, p. 115) points out, most researchers and theorists have chosen instead to suppress such overlap, or lessen its impact.

The basic model of ADCLUS has a familiar form:

$$s_{ij} \simeq \sum_k w_k p_{ik} p_{jk}$$

where

$p_{ik} = 1$ if object i has property k, and is 0 otherwise;
w_k is the (non-negative) weight representing the salience.

Intuitively, the model states that the similarity of any two objects i and j is the sum of the weights or saliences of the discrete properties which both objects have in common (Shepard 1980, p. 397). If both objects are in a given subset (i.e. share the same property) k, then $p_{ik} p_{jk} = 1$, but if either or both lie outside the subset the product becomes zero. So the similarity value equals the sum of the weights of just those clusters to which the objects belong. The weights w_k (often termed the elevation or salience weights in the literature) are interpreted as the importance of property k, and only contribute to the similarity of those objects which possess it.

The model has an obvious resemblance to the three-way CANDECOMP except that there are no individual subject effects—only one for each subset k. If the weight is ignored, the model reduces to a kind of binary scalar products (factor) model.

The estimation of the ADCLUS model presents problems of considerable complexity—not least that of information explosion. To find a non-hierarchical clustering of a set of p objects which best represent a lower-triangular set of $p(p - 1)/2$ (dis)similarity measures involves searching through $\{2^{(2^{p} - 1)} - 1\}$ distinct ways in which the objects might be clustered. For five objects this is over 2,147 million, and for eight objects it exceeds 10^{76}—clearly an impossibly high and totally unfeasible number of searches. The method in fact adopted is an iterative, alternating least squares procedure, which can use the researcher's 'guesstimate' of the likely subsets as a starting point. The procedure is described in detail in Arabie and Carroll (1980).

As in the case of additive tree analysis, it is useful to represent an additive clustering in a two-dimensional space obtained by scaling the data separately, and to do so by drawing circles around the subsets in a manner akin to HCS—except that for ADCLUS the circles will obviously intersect.

So far, applications of ADCLUS are, like ADDTREE, restricted principally to the methodological area. But they are interesting and exciting in their potential. The simplest example, based upon data from Shepard et al. (1975), took an aggregated lower-triangular matrix of judged similarities between the integers 0, 1, ..., 9. The solution for the first 10 subsets (accounting for 83 per cent of the data by the 10 w_k parameters) was as follows (Shepard and Arabie 1979, p. 103):

Rank	Weight (w_k)	Elements of Cluster	Interpretation of Cluster
1	0.577	{2, 4, 8}	*Powers of 2
2	0.326	{6, 7, 8, 9}	Large numbers
3	0.305	{3, 4, 5, 6}	Middling numbers
4	0.299	{1, 2, 3}	Small non-zero numbers
5	0.277	{3, 6, 9}	*Multiples of 3
6	0.165	{0, 1}	*Null and identity
7	0.150	{1, 3, 5, 7, 9}	Odd numbers
8	0.138	{5, 6, 7}	Moderately large numbers
9	0.112	{0, 1, 2}	Small numbers
10	0.101	{0, 1, 2, 3, 4}	Smallish numbers

(The clusters marked with an asterisk were detected using HCS)

Other applications of ADCLUS presented in the basic literature include the following:

(a)   Reported in Shepard and Arabie 1979

Brief description	Data reference
*1. Confusions between 16 consonant phonemes	Miller-Nicely
2. Confusions between 26 upper-case letters	Gibson
3. Correlations between stacked sociometric matrices for 14 Hawthorne Study workers	Roethlisberger and Dickson (and Breiger et al. 1975)
4. Judged similarity between 20 body parts (reproduced in Figure 8.2)	Miller (and Carroll and Chang 1973)

*(Note   The Miller-Nicely data are re-analysed using MAPCLUS in Arabie and Carroll 1980.)

(b)   In Arabie (1977)

5. Sociometric interaction between 14 biomedical researchers	Griffith (and Breiger et al. 1975)

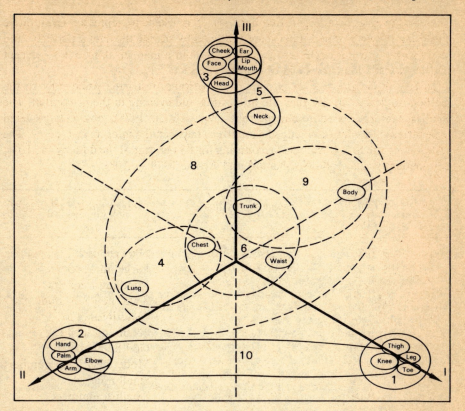

Figure 8.2   *The 10 ADCLUS subset obtained for the 20 body parts studied by Miller, embedded in a 3-dimensional scaling representation*

It seems clear that ADCLUS is set to be an important and increasingly used model in the future. Extensions to three-way models are also under development, with the predictable acronym of INDCLUS, for INDividual difference additive CLUStering (Carroll and Arabie 1979). In this case each subject i has a set of weights $w_k^{(i)}$ and the model becomes even more obviously akin to the scalar products version of INDSCAL.

8.5   Finale

Multidimensional scaling was identified in the first chapter as a family of models for the description and representation of data and the processes generating them. It should now be clear that the family is complex and growing. What began as an interesting but highly restricted attempt to reduce the dimensionality for reproducing the adjudged psychological similarities has become as extensive as the general linear model—and, indeed, has merged with it at certain points. Nonetheless, the basic framework developed here should suffice for encompassing newer developments and motivate researchers to adapt and use them in their particular fields of interest.

References and Bibliography

Peferences preceded by an asterisk (*) are reprinted in the accompanying volume of readings: P.M. Davies and A.P.M. Coxon, eds (1982), Key texts on multidimensional scaling, London: Heinemann.

Abelson, R. P. and John W. Tukey (1959), 'Efficient conversion of non-metric information into metric information', *American Statistical Association*, Social Statistics Section, Washington D.C., December 1959.

Adams, E. W. (1966), 'On the nature and purpose of measurement', Technical Report No. 4, University of Oregon, reprinted in B. Lieberman (ed.), *Contemporary problems in statistics*, Oxford: University Press, 74–91.

Alt, J., Bo Sarlvick and Ivor Crewe (1976), 'Individual differences scaling and group attitude structure: British party imagery in 1974', *Quality and Quantity*, **10**, 297–320.

Andrews, D. F. (1972), 'Plots of high-dimensional data', *Biometrics*, **28**, 125–36.

Anglin, J. A. (1970), *The growth of word meaning*, London: MIT Press.

Arabie, P. (1973), 'Concerning Monte Carlo evaluations of non-metric multidimensional scaling algorithms', *Psychometrika*, **38**, 607–8.

Arabie, P. (1977), 'Clustering representations of group overlap', *Journal of Mathematical Sociology*, **5**, 113–28.

Arabie, P. and S. A. Boorman (1973), 'Multidimensional scaling of measures of distance between partitions', *Journal of Mathematical Psychology*, **10**, 148–203.

Arabie, P. and J. D. Carroll (1980), 'MAPCLUS: a mathematical programming approach to fitting the ADCLUS model', *Psychometrika*, **45**, 211–35.

Arabie, P. and S. D. Soli (1977), 'The interface between the type of regression and methods of collecting proximities', in Golledge and Rayner 1981 (q.v.).

Arnold, J. B. (1971), 'A multidimensional scaling study of semantic distance', *Journal of Experimental Psychology*, **90**, 349–72.

Attneave, F. A. (1950), 'Dimensions of similarity', *American Journal of Psychology*, **63**, 516–56.

Bannister, D. and J. Mair (1969), *The evaluation of personal constructs*, London: Academic Press.

Barnett, G. A. (1977), 'Bilingual semantic organization: a multidimensional analysis', *Journal of Cross-cultural Psychology*, **8**, 315–30.

Beals, R., D. H. Krantz and A. Tversky (1968), 'Foundations of multidimensional scaling', *Psychological Review*, **75**, 127–42.

Bell Bibliography of Multidimensional Scaling Programs (1973), developed at the University of Michigan, Bell Telephone Laboratories and the University of North Carolina, including condensed description and abstracts of unpublished programs.

Berzécri, J. P. (1964), 'Sur l'analyse factorielle des proximites', *Publications of the Institute of Statistics, University of Paris*, **13**, 235–81.

Berent, J. (1952), 'Fertility and social mobility', *Population Studies*, **5**, 224–60.

Bick, W., P. J. Mueller, H. Bauer and O. Gieseke (1977), *Multidimensional scaling and clustering techniques (theory and applications in the social sciences) – a bibliography*, Cologne: Institute for Applied Social Research.

Blalock, H. M. (1972), *Social Statistics*, 2nd ed., New York: McGraw Hill.

Blau, P. M. and O. D. Duncan (1967), *The American occupational structure*, New York: Wiley.

Bloxom, B. (1974), 'An alternative method of fitting a model of individual differences in multidimensional scaling', *Psychometrika*, **39**, 365–67.

Boorman, S. A. and D. C. Oliver (1973), 'Metrics on spaces of finite trees', *Journal of Mathematical Psychology*, **10**, 26–59.

*Borg, I. (1977), 'Some basic concepts of facet theory', in Lingoes 1977, and 1979, 65–102.

Borg, I. (1980), 'Geometric representation of individual differences', in Lingoes, Roskam and Borg (1980), 609–56.

*Borg, I. and J. C. Lingoes (1978), 'What weight should weights have in individual differences scaling?' *Quality and Quantity*, **12**, 223–37.

Borg, I. and J. C. Lingoes (1980), 'A model and algorithm for multidimensional scaling with external constraints on the distances', *Psychometrika*, **45**, 25–38.

Box, G. E. P. and D. R. Cox (1964), 'An analysis of transformations', *Journal of the Royal Statistical Society* (*B*), **26**, 211–52.

Breiger, R. L., S. A. Boorman and P. Arabie (1975), 'An algorithm for clustering relational data with applications to social network analysis and comparison with multidimensional scaling', *Journal of Mathematical Psychology*, **12**, 328–82.

Buneman, P. (1971), 'The recovery of trees from measures of dissimilarity', in Hodson *et al.* (1971) (q.v.).

Burton, M. L. (1972), 'Semantic dimensions of occupation names', in Romney, Shepard and Nerlove 1972 (q.v.).

Burton, M. L. (1975), 'Dissimilarity measures for unconstrained sorting data', *Multivariate Behavioral Research*, **10**, 409–24.

Burton, M. L. and S. B. Nerlove (1976), 'Balanced designs for triads tests: two examples from English', *Social Science Research*, **5**, 247–67.

Burton, M. L. and A. K. Romney (1975), 'A multidimensional representation of role terms', *American Ethnologist*, **2**, 397–407.

Canter, D. and S. K. Tagg (1975), 'Distance estimation in cities', *Environment and Behaviour*, **7**, 59–80.

Carroll, J. D. (1970), Book review of Delbeke 1968, *Psychometrika*, **35**, 278–81.

*Carroll, J. D. (1972), 'Individual differences and multidimensional scaling', in Shepard, Romney and Nerlove 1972 (q.v.), 105–55.

Carroll, J. D. (1973), 'Models and algorithms for multidimensional scaling, conjoint measurement, and related techniques', in Paul E. Green, and Yoram Wind, *Multiattribute decisions in marketing: a measurement approach*, New York: The Dryden Press, 300–96.

Carroll, J. D. (1975), 'Application of CANDECOMP to solving for parameters of Lazarsfeld's latent class model'. Mimeo—Handout of Society for Multivariate Experimental Psychology meeting.

Carroll, J. D. (1976) 'Spatial, non-spatial and hybrid models for scaling', *Psychometrika*, **41**, 439–63.

Carroll, J. D. and P. Arabie (1979), 'INDCLUS', Monterey: Psychometric Society, mimeo.

Carroll, J. D. and P. Arabie (1980), 'Multidimensional scaling', in M. R. Rosenzweig and L. W. Porter, eds. *Annual Review of Psychology*, Palo Alto: Annual Reviews.

Carroll, J. D. and J-J. Chang (1967), 'Relating preference data to multidimensional scaling solutions via a generalization of Coombs' unfolding model', Murray Hill: Bell Laboratories.

*Carroll, J. D. and J-J. Chang (1970), 'Analysis of individual differences in multidimensional scaling via an N-way generalization of Eckart-Young decomposition', *Psychometrika*, **35**, 283–319.

Carroll, J. D. and J-J. Chang (1971), 'An alternative solution to the "metric unfolding" problem'. Paper to Psychometric Society at St. Louis, Missouri.

Carroll, J. D. and J-J. Chang (1973), 'A method for fitting a class of hierarchical tree structure models to dissimilar data and its application to some "body parts" data of

Miller's', *Proceedings*: 81st Annual Convention of the American Psychological Association, 197–8.

Carroll, J. D. and P. E. Green (1976), 'CANDELINC (CANonical DEcomposition with LINear Constraints): A new method of multidimensional analysis with constrained solutions'. Mimeo.

Carroll, J. D. and F. Pruzansky (1975), 'Fitting of hierchical tree structures (HTS) models, mixture of HTS models, and hybrid models, via mathematical programming and alternating least squares'. Paper presented at the US-Japan Seminar on Theory, Methods, and Applications of Multidimensional Scaling and Related Techniques.

Carroll, J. D. and M. Wish (1973), 'Models and methods for three-way multidimensional scaling', in *Contemporary developments in mathematical psychology*, ed. by R. C. Atkinson, D. H. Krantz, R. D. Luce and P. Suppes, San Francisco: W. H. Freeman.

*Carroll, J. D. and M. Wish (1974), 'Multidimensional perceptual models and measurement of methods', in E. C. Carterette and M. P. Friedman (eds.), *Handbook of Perception, Vol. II*, New York: Academic Press, 391–447.

Carterette, E. C. and M. P. Friedman (eds.) (1974), *Handbook of Perception, Vol. II*, New York: Academic Press.

Cliff, N. (1968), 'Adjective check list responses and individual differences in perceived meaning', *Educational and Psychological Measurement*, **28**, 1,063–77.

Cochran, W. G. and G. M. Cox (1951), *Experimental design*, New York: John Wiley.

Coombs, C. H. (1950), 'Psychological scaling without a unit of measurement', *Psychological Review*, **57**, 145–58.

Coombs, C. H. (1964), *A theory of data*, New York: John Wiley.

Coombs, C. H. (1967), 'Thurstone's measurement of social values revisited forty years later', *Journal of Personality and Social Psychology*, **6**, 85–91.

Coombs, C. H., R. M. Dawes and A. Tversky (1970), *Mathematical psychology: an elementary introduction*, Englewood Cliffs: Prentice Hall.

Cooper, L. G. (1971), 'Procedure for metric multidimensional scaling', reprinted from the Proceedings, 79th Annual Convention, APA 1971, 87–8.

Cormack, R. M. (1971), 'A review of classification', *Journal of the Royal Statistical Society (A)*, **134**, 321–67.

Coxon, A. P. M. (1971), 'Occupational attributes: constructs and structure', *Sociology*, **5**, 335–54.

*Coxon, A. P. M. (1974), 'The mapping of family-composition preferences: a scaling analysis', *Social Science Research*, **3**, 191–210.

Coxon, A. P. M. and P. M. Davies (1979), *Working papers in multidimensional scaling*, Proc. MDS(X) Seminar, University of Cambridge, September 1979, mimeo, Cardiff: MDS(X) project.

Coxon, A. P. M. and C. L. Jones (1977), 'Multidimensional scaling' (ch. 6), *Analysis of survey data; vol. I: exploring data structures*, C. Payne and C. A. O'Muircheartaigh (q.v.).

Coxon, A. P. M. and C. L. Jones (1978a), *The images of occupational prestige: a study in social cognition*, London: Macmillan.

Coxon, A. P. M. and C. L. Jones (1978b), *Class and hierarchy: the social meaning of occupations*, London: Macmillan.

Coxon, A. P. M. and C. L. Jones (1979), *Measurement and meanings: techniques and methods of studying occupational cognition*, London: Macmillan.

Coxon, A. P. M. and C. L. Jones (1980), 'Multidimensional scaling: exploration to confirmation', *Quality and Quantity*, **14**, 31–74.

Cronbach, L. J. (1946), 'Response sets and test validity', *Educational Psychological Measurement*, **6**, 475–94.

Cunningham, J. P. (1978), 'Free trees and bidirectional trees as representations of psychological distance', *Journal of Mathematical Psychology*, **17**, 165–88.

Cunningham, J. P. and R. N. Shepard (1974), 'Monotone mapping of similarities into a general metric space', *Journal of Mathematical Psychology*, **11**, 335–63.

D'Andrade, R. G. (1976), 'A propositional analysis of US American beliefs about illness', in K. H. Basso and H. A. Selby (eds.), *Meaning in anthropology*, Albuquerque: University of New Mexico Press, 155–80.

David, H. A. (1963), *The method of pair comparisons*, London: Griffin.

Davison, M. L. (1976), 'Fitting and testing Carroll's weighted unfolding model for preferences', *Psychometrika*, **41**, 233–47.

Davison, M. L. (1980), 'DACAR: A Fortran IV program for fitting linear and distance models of preference', Minneapolis: University of Minnesota, mimeo.

Davison, M. L. and L. E. Jones (1976), 'A similarity-attraction model for predicting sociometric choice from perceived group structure', *Journal of Personality and Social Psychology*, **33**, 601–12.

Dawes, R. M. (1972), *Fundamentals of attitude measurement*, New York: Wiley.

Delbeke, L. (1968), *Construction of preference spaces*, Louvain: University Press.

Doyle, P. and P. Hutchinson (1973), 'Individual differences in family decision making', *Journal of the Market Research Society*, **15**, 281–8.

Durbin, J. (1951), 'Incomplete blocks in ranking experiments', *British Journal of Psychology, Statistical Section*, **4**, 85–90.

Eckart, C. and G. Young (1936), 'Approximation of one matrix by another of lower rank', *Psychometrika*, **1**, 211–18.

Ellis, B. (1966), *Basic concepts of measurement*, Cambridge: University Press.

Erlandson, R. F. (1978), 'System evaluation methodologies: combined MDS and ordering techniques', *IEEE Transactions on systems, man and cybernetics, SMC 8*, 421–32.

Everitt, B. (1974), *Cluster analysis*, London: Heinemann Educational Books.

Everitt, B. (1978), *Graphical techniques for multivariate data*, London: Heinemann Educational Books.

Fielding, A. (1977), 'Latent structure models', in O'Muircheartaigh and Payne (q.v.) 1972.

Fillenbaum, S. and A. Rapoport (1971), *Structure in the subjective lexicon*, New York: Academic Press.

Flament, C. (1963), *Applications of graph theory to group structure*, Englewood Cliffs: Prentice Hall.

*Forgas, J. P. (1978), 'Multidimensional scaling: a discovery method in social psychology', in G. P. Ginsburg (ed.), (1978), *Emerging strategies in social psychological research*, London: Wiley.

Friendly, M. L. and S. Glucksberg (1970), 'On the description of subcultural lexicons: a multidimensional approach, *Journal of Personality and Social Psychology*, **14**, 55–65.

Gabrielsson, A. (1973), 'Similarity ratings and dimension analyses of auditory rhythm patterns', *Scandinavian Journal of Psychology*, **14**, 138–60 and 161–76.

Galtung, J. (1967), *Theory and methods of social research*, London: Allen and Unwin.

Giles, H., D. Taylor and R. Bowhis (1977), 'Dimensions of Welsh identity', *European Journal of Social Psychology*, **7**, 165–74.

Gleason, T. C. (1969), 'Multidimensional scaling of sociometric data', The University of Michigan, Ph.D. Thesis.

Gnanadesikan, R. (1973), 'Graphical methods for informal inference in multivariate data analysis', *Bulletin of the International Statistical Institute*, Proc. of 39th session, 195–206.

Golledge, R. G. (1973), 'On determining cognitive configurations of a city', vol. 1, Ohio: Department of Geography, Ohio State University, mimeo.

Golledge, R. G. and J. N. Rayner (eds.) (1981), *Multidimensional analysis of large data sets*, Columbus: Ohio State University Press.

Goodman, L. A. and W. H. Kruskal (1954), 'Measures of association for cross-classifications', *Journal of the American Statistical Association*, **49**, 732–64 and **54**, 123–63 and **58**, 310–64.

Gower, J. C. (1975), 'Generalised procrustes analysis', *Psychometrika*, **40**, 33–51.

Gower, J. C. (1976), 'The analysis of three-way grids', in P. Slater (ed.), *The measurement of intrapersonal space*, London: Wiley.

Gower, J. C. (1977), 'The analysis of asymmetry and orthogonality', in J. R. Barra et al., (eds.), *Recent developments in statistics*, Amsterdam: North-Holland.

Gower, J. C. (1979), 'Comparing multidimensional scaling configurations', in Coxon and Davies 1979 (q.v.).

Gower, J. C. (1980), 'Some characterisations of matrix multidimensional scaling methods', *Journal of the Royal Statistical Society*, in press.

Graef, J. and I. Spence (1979), 'Using distance information in the design of large multidimensional scaling experiments', *Psychological Bulletin*, **86**, 60–6.

Grasmick, H. G. (1976), 'The occupational prestige structure: a multidimensional scaling approach', *The Sociological Quarterly*, **17**, 90–108.

Green, P. E. and F. J. Carmone (1970), *Multidimensional scaling and related techniques in marketing analysis*, Boston: Allyn & Bacon.

Green, P. E. and F. J. Carmone (1971), 'Stimulus context and task effects on individuals' similarities judgements', in *Attitude research reaches new heights*, C. W. King and D. Tigert (eds.), Chicago: American-Marketing Assoc.

Green, P. E. and J. D. Carroll (1976), *Mathematical tools for applied multivariate analysis*, New York: Academic Press.

Green, P. E. and V. R. Rao (1970), 'Rating scales and information recovery—how many scales and response categories to use?' *Journal of Marketing*, **34**, 33–9.

Green, P. E. and V. R. Rao (1972), *Applied multidimensional scaling: a comparison of approaches and algorithms*, New York: Holt, Rinehart.

Green, P. E. and Y. Wind (1975), 'New way to measure consumers' judgements', *Harvard Business Review*, 107–17.

Gregson, R. A. M. (1966), 'Theoretical and empirical multidimensional scalings of taste mixture matchings', *British Journal of Mathematical and Statistical Psychology*, **18**, 59–75.

Guttman, L. (1968), 'A general non-metric technique for finding the smallest co-ordinate space for a configuration of points', *Psychometrika*, **33**, 469–506.

Guttman, L. (1971), 'Measurement as structural theory', *Psychometrika*, **36**, 329–47.

Harshman, R. A. (1972), 'PARAFAC 2: mathematical and technical notes', Los Angeles (UCLA) Working Papers in Phonetics 22.

Hartman, M. (1979), 'Prestige grading of occupations with sociologists as judges', *Quality and Quantity*, **13**, 1–21.

Hayashi, C. (1968), 'One-dimensional quantification and multidimensional quantification', *Annals of the Japanese Association for the Philosophy of Science*, **3**, 115–19.

Heider, E. R. and D. C. Olivier (1972), 'The structure of the color space in naming and memory for two languages', *Cognitive Psychology*, **3**, 337–54.

Heiser, W. and J. de Leeuw (1979), 'Multidimensional mapping of preference data', Leiden: Dept. of Data theory, mimeo.

Hodson, F. R., D. G. Kendall and P. Tautu (eds.) (1971), *Mathematics in the archaeological and historical sciences*, Edinburgh: University Press.

Holman, E. W. (1972), 'The relation between hierarchical and Euclidean models for psychological distances', *Psychometrika*, **37**, 417–23.

Hope, K. (ed.), (1972), *The analysis of social mobility: methods and approaches*, Oxford: Clarendon Press.

Horan, C. B. (1969), 'Multidimensional scaling: combining observations when individuals have different perceptual structures', *Psychometrika*, **34**, 139–65.

Hubert, L. (1974), 'Some applications of graph theory and related non-metric techniques to problems of approximate seriation: the case of symmetric proximity measures', *British Journal of Mathematical and Statistical Psychology*, **27**, 133–53.

Hyman, R. and A. Well (1968), 'Perceptual separability and spatial models', *Perception and Psychophysics*, **3**, 161–65.

Inoguchi, T. (1970), 'Attitudes toward war and peace: a quantification scaling analysis', Tokyo: Sophia University, Institute of International Relations, Research Papers Series A-6.

Isaac, P. D. and D. D. S. Poor (1974), 'On the determination of appropriate dimensionality in data with error', *Psychometrika*, **39**, 91–109.

Jain, A. K. and M. Etgar (1976), 'Measuring store image through MDS of free response data', *Journal of Retailing*, **52**, 61–96.

Jardine, N. and R. Sibson (1971), *Mathematical taxonomy*, London: Wiley.

Johnson, S. C. (1967), 'Hierarchical clustering schemes', *Psychometrika*, **32**, 241–54.

Jones, L. E. and J. Waddington (1973), 'Sensitivity of INDSCAL to simulated individual differences in dimension usage patterns and judgemental error', Urbana: University of Illinois, mimeo.

Jones, L. E. and F. W. Young (1972), 'Structure of a social environment: A longitudinal individual differences scaling of an intact social group, *Journal of Personality and Social Psychology*, **24**, 108–21.

Jones, R. A. and R. D. Ashmore (1973), 'The structure of intergroup perception: categories and dimensions in views of ethnic groups and adjectives used in stereotype research', *Journal of Personality and Social Psychology*, **25**, 428–38.

Kelly, G. (1955), *The psychology of personal constructs*, New York: Norton.

Kendall, D. G. (1962), *Rank correlation methods*, 3rd ed., London: Griffin.

Kendall, D. G. (1971a), 'Seriation from abundance matrices', in Hodson, et al. (eds.) 1971 (q.v.), 215–52.

*Kendall, D. G. (1971b), 'Maps from marriages: an application of non-metric multidimensional scaling to parish register data', in Hodson et al., 1971, 303–18.

Kendall, D. G. (1971c), 'Construction of maps from "odd bits of information"', *Nature*, **231**, 158–9.

Kirk, G. S. (1974), *The nature of Greek myths*, Harmondsworth: Penguin Books.

Klahr, D. (1969), 'A Monte Carlo investigation of the statistical significance of Kruskal's nonmetric scaling procedure', *Psychometrika*, **34**, 319–31.

Koopman, R. F. and M. Cooper (1974), 'Some problems with Minkowski distance models in multidimensional scaling'. Paper presented to the Psychometric Society, 28 March 1974, Stanford, Ca. Mimeo.

Krantz, D. H., R. D. Luce, P. Suppes and A. Tversky (1971), *Foundations of measurement. Volume 1: Additive and polynomial representations*, New York: Academic Press.

*Kruskal, J. B. (1964a), 'Multidimensional scaling by optimizing goodness of fit to a nonmetric hypothesis', *Psychometrika*, **29**, 1:1–27.

*Kruskal, J. B. (1964b), 'Nonmetric multidimensional scaling: a numerical method', *Psychometrika*, **29**, 115–29.

Kruskal, J. B. (1965), 'Analysis of factorial experiments by estimating monotone transformations of the data', *The Journal of the Royal Statistical Society*, **27**, 251–63.

Kruskal, J. B. (1972a), 'Statistical aspects of scaling'. Mimeo. MDS Workshop.

Kruskal, J. B. (1972b), 'Special problems and possibilities'. Mimeo. MDS Workshop.

Kruskal, J. B. and J. D. Carroll (1969), 'Geometrical models and badness-of-fit functions', in P. R. Krishnaiah (ed.): *Multivariate analysis, II*, New York: Academic Press, 1969.

Kruskal, J. B. and R. E. Hart (1966), 'A geometric interpretation of diagnostic data from a digital machine: based on a study of the Morris, Illinois Electronic Central Office', *The Bell System Technical Journal*, **45**, 1,299–338.

Kruskal, J. B. and M. Wish (1978), *Multidimensional scaling*, (Sage University Paper series on Quantitative Application in the Social Sciences, series No. 07–011), London: Sage Publications.

Kruskal, J. B., F. W. Young and J. B. Seery (1973), 'How to use KYST, a very flexible program to do multidimensional scaling and unfolding, Murray Hill: Bell Laboratories, mimeo.

Langeheine, R. (1980), 'Approximate norms and significance tests for the LINGOES-BORG procrustes individual differences scaling (PINDIS)', Kiel: Institut für die Pädagogik der Naturwissenschaften (IPN), mimeo.

Laumann, E. O. (1973), *Bonds of pluralism: the form and substance of urban social networks*, London: John Wiley.

Laumann, E. O. and L. Guttman (1966), 'The relative associational contiguity of occupations in an urban setting', *American Sociological Review*, **31**, 169–78.

Lazarsfeld, P. F. and N. W. Henry (1968), *Latent structure analysis*, New York: Houghton Mifflin.

Levelt, W. J. M., J. P. van de Geer and R. Plomp (1966), 'Triadic comparisons of musical intervals', *British Journal of Mathematical and Statistical Psychology*. **19** (part 2), 163–79.

Levine, J. H. (1972), 'The sphere of influence', *American Sociological Review*, **37**, 14–27.

Levy, S. (1976), 'Use of the mapping sentence for coordinating theory and research: a cross-cultural example', *Quality and Quantity*, **10**, 117–25.

Levy, S. and L. Guttman (1975), 'On the multivariate structure of well-being', *Social Indicators Research*, **2**, 361–88.

Lingoes, J. C. (1977), 'Progressively complex linear transformations for finding geometric similarities among data structures'. Mimeo.

Lingoes, J. C. (1979), 'Identifying regions in the space for interpretation', in Lingoes et al', q.v. (eds.), 1979, 115–26.

Lingoes, J. C. and I. Borg (1978), 'A direct approach to individual differences scaling using increasingly complex transformations', *Psychometrika*, **43**, 491–519.

Lingoes, J. C. and I. Borg (1979), 'Identifying spatial manifolds for interpretation, in Lingoes et al. (eds.), 1979, 127–47.

Lingoes, J. C. and E. E. Roskam (1973), 'A mathematical and empirical analysis of two multidimensional scaling algorithms', *Psychometrika*, **38** (4, pt. 2, monograph supplement).

Lingoes, J. C., E. E. Roskam and I. Borg (eds.) (1979 and 1980), *Geometric representation of relational data*, Ann Arbor: Mathesis Press.

Lingoes, J. C. and P. H. Schönemann (1974), 'Alternative measures of fit for the Schönemann-Carroll matrix fitting algorithm', *Psychometrika*, **39**, 423–7.

Loether, H. J. and D. G. McTavish (1974), *Descriptive statistics for sociologists*, Boston: Allyn & Bacon.

Luce, R. D. (1956), 'Semi-orders and theory of utility discrimination', *Econometrica*, **24**, 158–67.

Lundberg, U., O. Bratfisch and G. Ekman (1972), 'Emotional involvement and subjective distance: a summary of investigations', *Journal of Social Psychology*, **87**, 169–77.

Lundberg, U. and G. Ekman (1973), 'Subjective geographic distance: a multidimensional comparison', *Psychometrika*, **38**, 113–22.

MacCallum, R. C. (1977), 'Effects of conditionality on INDSCAL and ALSCAL weights', *Psychometrika*, **42**, 297–305.

Macdonald, K. I. (1972), MDSCAL and distances between socio-economic groups, in K. Hope (ed.): *The analysis of social mobility*, Oxford: Clarendon Press, 141–64.

Maimon, Z. (1978), 'The choice of ordinal measures of association', *Quality and Quantity*, **12**, 255–64.

Maimon, Z., I. Venezia and J. C. Lingoes (1980), 'How similar are the different results?' *Quality and Quantity*, **14**, 727–43.

Mardia, K. V. (1972), *Statistics of directional data*, London: Academic Press.

Maxwell, A. E. (1977), *Multivariate analysis in behavioural research*, London: Chapman and Hall.

MDS(X) User's Manual (SV3) (1980) Edinburgh: Program Library Unit.

Messick, S. J. and R. P. Abelson (1956), 'The additive constant problem in multidimensional scaling, *Psychometrika*, **21**, 1–15.

Miller, G. A. (1956), 'The magical number seven, plus or minus two: some limits to our capacity for processing information', *Psychological Review*, **63**, 81–97.

Miller, G. A. (1969), 'Psychological method to investigate verbal concepts, *Journal of Mathematical Psychology*, **6**, 169–91.

Nabokov, V. V. (1955), *Lolita*, Paris: The Olympia Press.

Newcomb, T. M. (1961), *The acquaintance process*, New York: Holt, Rinehart.

O'Grady, K., L. H. Janda and H. B. Gillen (1979), 'A multidimensional scaling analysis of sex guilt', *Multivariate Behavioral Research*, **14**, 415–34.

O'Hare, D. (1976), 'Individual differences in perceived similarity and preference for visual art: a multidimensional scaling analysis', *Perception and Psychophysics*, **20**, 445–52.

O'Muircheartaigh, C. A. and C. Payne (eds.) (1977), *The analysis of survey data vol. I: Exploring data structures*, London: Wiley.

Osgood, C. E., G. J. Suci and P. H. Tannenbaum (1965), 'The measurement of meaning', Urbana: University of Illinois Press.

Preece, P. F. W. (1976), 'Science concepts in semantic space—a MDS study', *Alberta Journal of Educational Research*, **22**, 281–88.

Prim, R. C. (1957), 'Shortest connection networks and some generalizations', *The Bell System Technical Journal*, **36**, 1,389–1,607.

*Ramsay, J. O. (1977), 'Maximum likelihood estimation in MDS', *Psychometrika*, **42**, 241–66.

Rao, V. R. and R. Katz (1971), 'Alternative multidimensional scaling methods for large stimulus sets', *Journal of Marketing Research*, **8**, 488–94.

Rasch, G. (1960), *Probabilistic models for some intelligence and attainment tests*, Copenhagen: Nielsen and Lydiche.

Reeb, M. (1959), 'How people see jobs: a multidimensional analysis', *Occupational Psychology*, **33**, 1–17.

Renfrew, C. (1976), *Before civilization*, Harmondsworth: Penguin.

Rescher, N. (1969), *Introduction to value theory*, Englewood Cliffs NJ: Prentice Hall.

Restle, F. (1959), 'A metric and an ordering on sets', *Psychometrika*, **24**, 207–20.

Retka, R. L. and R. M. Fenker Jr. (1975), 'Self-perception among narcotic addicts: an exploratory study employing multidimensional scaling techniques', *The International Journal of the Addictions*, **10**, 1–12.

Richards, L. G. (1972), 'A multidimensional scaling analysis of judged similarity of complex forms from two task situations', *Perception and Psychophysics*, **12**(2A), 154–160.

Richardson, M. W. (1938), 'Multidimensional psychophysics', *Psychological Bulletin*, **35**, 659–60.

Rivett, P. (1977), 'MDS for multiobjective policies', *Omega*, **5**, 367–79.

Rivizzigno, V. L. (1973), 'Individual differences in the cognitive structuring of an urban area', in Golledge, 1973 (q.v.).

Robertson, C. M. (1976), 'Individual differences in perception of sex by Siamese fighting fish', unpublished manuscript; reported in Spence 1978 (q.v.), 201 et seq.

Robinson, J. P. and R. Hefner (1967), 'Multidimensional differences in public and academic perceptions of nations', *Journal of Personality and Social Psychology*, **7**, 251–59.

Romney, A. K., R. N. Shepard and S. B. Nerlove (eds.) (1972), *Multidimensional scaling: theory and applications in the behavioral sciences, Vol. II: Applications*, New York: Seminar Press.

Rorer, L. G. (1965), 'The great response-style myth', *Psychological Bulletin*, **63**, 129–56.

Rosenberg, S. and R. Jones (1972), 'A method for investigating and representing a person's implicit theory of personality: Theodore Dreiser's view of people', *Journal of Personality and Social Psychology*, **22**, 372–86.

Rosenberg, S. and A. Sedlak (1972a), 'Structural representations of perceived personality trait relationships', in Romney et al, (eds.) 1974, 134–63.

Rosenberg, S. and A. Sedlak (1972b), 'Structural representations of implicit personality theory', in *Advances in experimental social psychology, vol. 6*, New York & London: Academic Press, Inc., 235–97.

Roskam, E. E. Ch.I. (1968), *Metric analysis of ordinal data in psychology*, Nijmegen: Voorschoten.

Roskam, E. E. Ch.I. (1969), 'A comparison of principles for algorithm construction in nonmetric scaling', Ann Arbor: MMPP Report 69-2.

Roskam, E. E. (1970), 'The method of triads for nonmetric MDS', *Psychologie*, **25**, 404–17.

Roskam, E. E. (1972), 'MDS by metric transformation of data', *Psychologie*, **27**, 486–508.

Roskam, E. E. (1975), 'Nonmetric data analysis: general methodology and technique with brief descriptions of mini-programs'. Dept. of Psychology, University of Nijmegen, Report 75-MA-13.

Roskam, E. E. (1979), 'A survey of the Michigan-Israel-Netherlands-integrated series', in Lingoes et al, (eds.) 1979 (q.v.), 289–312.

Ruch, L. O. (1977), 'A multidimensional analysis of the concept of life change', *Journal of Health and Social Behaviour*, **18**, 71–83.

Rummel, R. J. (1970), *Applied factor analysis*, Evanston: North Western University Press.

Sattath, S. and A. Tversky (1977), 'Additive similarity trees', *Psychometrika*, **42**, 319–46.

Schiffman, S. S., M. L. Reynolds and F. W. Young (1982), *Introduction to multidimensional scaling: theory, method and applications*, London: Academic.

Schlesinger, I. M. and L. Guttman (1969), 'Smallest space analysis of intelligence and achievement tests', *Psychological Bulletin*, **71**, 95–100.

Schmidt, C. F. (1972), 'Multidimensional scaling analysis of the printed media's explanations of the riots of the summer of 1967', *Journal of Personality and Social Psychology*, **24**, 59–67.

Schneider, E. J. and H. F. Weisberg (1974), 'An interactive graphics approach to dimensional analysis', *Behavioral Research Methods and Instrumentation*, **6**, 185–94.

Schönemann, P. H. (1970), 'Metric multidimensional unfolding', *Psychometrika*, **35**, 349–66.

Schönemann, P. H. (1972), 'An algebraic solution for a class of subjective metrics models', *Psychometrika*, **37**, 441–51.

Seligson, M. A. (1977), 'Prestige among peasants: a multidimensional analysis of preference data', *American Journal of Sociology*, **83**, 632–52.

Shepard, R. N. (1962), 'The analysis of proximities: MDS with an unknown distance function I', *Psychometrika*, **27**, 125–40.

Shepard, R. N. (1962), 'The analysis of proximities: MDS with an unknown distance function II', *Psychometrika*, **27**, 219–45.

Shepard, R. N. (1963), 'Analysis of proximities as a technique for the study of information processing in man, *Human Factors*, **5**, 33–48.

*Shepard, R. N. (1966), 'Metric structures in ordinal data', *Journal of Mathematical Psychology*, **3**, 287–315.

Shepard, R. N. (1972a), '2-way MDS applications'. MDS Workshop. Mimeo.

Shepard, R. N. (1972b), 'Concepts and methods of MDS'. MDS Workshop: Introduction. Mimeo.

Shepard, R. N. (1972c), 'A taxonomy of some principle types of data and of multidimensional methods for their analysis', in Shepard, Romney and Nerlove, 1972 (q.v.), 21–47.

*Shepard, R. N. (1974), 'Representation of structure in similarity data: Problems and prospects', *Psychometrika*, **39**, 373–421.

Shepard, R. N. (1978), 'The circumplex and related topological manifolds in the study of perception', in Shye 1978 (q.v.), 29–80.

*Shepard, R. N. (1980), 'Multidimensional scaling, tree-fitting, and clustering', *Science*, **210** 390–98.

Shepard, R. N. and P. Arabie (1979), 'Additive clustering: representation of similarities as combinations of discrete overlapping properties', *Psychological Review*, **86**, 87–123.

Shepard, R. N. and J. D. Carroll (1966), 'Parametric representation of nonlinear data structures', in *Multivariate analysis*, P. N. Krishnaiah (ed.).

Shepard, R. N., D. W. Kilpatric and J. P. Cunningham (1975), 'The internal representation of numbers', *Cognitive Psychology*, **7**, 82–138.

Shepard, R. N., A. K. Romney and S. B. Nerlove (eds.) (1972), *Multidimensional scaling: theory and applications in the behavioural sciences, Vol. I: Theory*, New York: Seminar Press.

Sherman, R. C. and M. D. Dowdle (1974), 'The perception of crime and punishment: a multidimensional scaling analysis', *Soc. Sci. Res.*, **3**, 109–26.

Sherman, R. C. and L. B. Ross (1972), 'Liberalism-Conservatism and dimensional salience in the perception of political figures', *Journal of Personality and Social Psychology*, **23**, 120–27.

Shikiar, R. (1976), 'Multidimensional perceptions on the 1972 presidential election', *Multivariate Behavior Research*, **11**, 259–63.

Shye, S. (ed.) (1978), *Theory construction and data analysis in the behavioral sciences*, San Francisco: Jossey Bass.

Sibson, R. (1972), 'Order invariant methods for data analysis', *J. Royal Stat. Soc. (B)*, **34**, 311–49.

Sidowski, J. B. and N. H. Anderson (1967), 'Judgments of city-occupation combinations', *Psychon. Sci.*, **31**, 125–45.

Skinner, H. A. (1978), 'Differentiating the contribution of elevation, scatter and shape in profile similarity', *Educational and Psychological Measurement*, **38**, 297–308.

Slater, P. (1960), 'The analysis of personal preferences', *The British Journal of Statistical Psychology*, **13**, 119–35.

Sneath, P. H. A. and R. R. Sokal (1973), *Numerical taxonomy*, San Francisco: Freeman.

Sokal, R. R. and P. H. A. Sneath (1963), *Principles of numerical taxonomy*, San Francisco: Freeman.

Spence, I. (1972), 'A Monte Carlo evaluation of three nonmetric multidimensional scaling algorithms, *Psychometrika*, **37**, 461–86.

Spence, I. (1978), 'Multidimensional scaling', in *Quantitative ethology*, P. Colgan (ed.), New York: Wiley.

Spence, I. and D. W. Domoney (1974), 'Single subject incomplete designs for nonmetric multidimensional scaling', *Psychometrika*, **39**, 469–90.

Spence, I. and J. Graef (1974), 'The determination of the underlying dimensionality of an empirically obtained matrix of proximities', *Multivariate Behavioral Research*, **9**, 331–41.

Stenson, H. H. and R. L. Knoll (1969), 'Goodness of fit for random rankings in Kruskal's nonmetric scaling procedure', *Psychological Bulletin*, **71**, 122–26.

Stevens, S. S. (1946), 'On the theory of scales of measurement', *Science*, **103**, 677–80.

Stevens, S. S. (1959), 'Measurement, psychophysics and utility', in C. W. Churchman and P. Ratoosh (eds.), *Measurement: definition and theories*, New York: Wiley, 18–63.

Stevens, S. S. (1966), A metric for social consensus', *Science*, **151**, 530–1.

Stevens, S. S. (1974), 'Perceptual magnitude and its measurement, in Carterette and Friedman 1974 (q.v.) (ch.11).

Stewart, A., K. Prandy and R. M. Blackburn (1980), *Social stratification and occupations*, London: Macmillan.

Suppes, P. and M. Winet (1955), 'An axiomatization of utility based on the notion of utility differences', *Management Science*, **18**, 259–70.

Suppes, P. and I. L. Zinnes (1963), 'Basic measurement theory', in R. D. Luce, R. R. Bush and E. Galanter (eds.), *Handbook of mathematical psychology*, *Vol. 1*, New York: Wiley, 1–76.

Tagg, S. K. (1975), 'MDS in environmental psychology research: a summary review', *Architectural Psychology Newsletter*, **5**, 22–4.

Tagg, S. K. (1979), 'The analysis of repertory grids', in Coxon and Davies 1979 (q.v.).

Takane, Y., F. W. Young and J. de Leeuw (1977), 'Nonmetric individual differences MDS: An alternating least squares method with optimal scaling features', *Psychometrika*, **42**, 7–67.

Taylor, D. M., J. N. Bassili and F. E. Aboud (1973), 'Dimensions of ethnic identity: an example from Quebec, *The Journal of Social Psychology*, **89**, 185–92.

Thurstone, L. L. (1927), 'Method of paired comparisons for social values', *Journal of Abnormal and Social Psychology*, **21**, 384–400.

Tobler, W. and S. Wineberg (1971), 'A Cappadocian speculation', *Nature*, **231**, 39–41.

Torgerson, W. S. (1958), *Theory and methods of scaling*, New York: Wiley.

Traub, R. E. and R. K. Hambleton (1974), 'The effect of instruction on the cognitive structure of statistical and psychometric concepts', *Canadian Journal of Behavioral Science*, **6**, 30–44.

Tucker, L. R. (1964), 'Extension of factor analysis to three-dimensional matrices', in N. Frederiksen and H. Gulliksen (eds.), *Contributions to mathematical psychology*, New York: Holt, Rinehart & Winston, 109–27.

Tucker, L. R. (1972), 'Relations between multidimensional scaling and three-mode factor analysis', *Psychometrika*, **37**, 3–27.

Tversky, A. and D. H. Krantz (1970), 'Dimensional representation and the metric structure of similarity data', *Journal of Mathematical Psychology*, **7**, 572–96.

Tyler, S. A. (ed.) (1969), *Cognitive anthropology*, New York: Holt, Rinehart.

Ullrich, J. R. and M. F. Ullrich (1976), 'A multidimensional scaling analysis of perceived similarities of rivers in Western Montana', *Perceptual and Motor Skills*, **43**, 575–84.

van de Geer, J. P. (1971), *Introduction to multivariate analysis for the social sciences*, San Francisco: Freeman.

van Deth, J. (1979), 'PRO-FIT: a program for fitting properties in a multidimensional space using linear or nonlinear regression procedures', Amsterdam: Technisch Centrum Faculteit der Sociale Wetenschappen, TC-91.

van Schuur, W. H. (1977), 'The output from PREFMAP', Edinburgh: MDS(X) Report 8, IURCS Report No. 39.

Waern, Y. (1971), 'Structure in similarity matrices: a graphic approach', *Scandinavian Journal of Psychology*, **12**, 1–12.

Wagenaar, W. A. and P. Padmos (1971), 'Quantitative interpretation of stress in Kruskal's multidimensional scaling technique', *British Journal of Mathematical and Statistical Psychology*, **24**, 101–10.

Wallis, R. and R. Bland (1979), 'Purity in danger', *British Journal of Sociology*, **30**, 188–205.

Ward, L. M. (1977), 'MDS of the molar physical environment, *Multivariate Behavioral Research*, **12**, 23–42.

Warr, P. B., H. M. Schroder and S. Blackman (1969), 'A comparison of two techniques for the measurement of international judgment', *International Journal of Psychology*, **4**, 135–140.

Weisberg, H. F. and J. G. Rusk (1970), 'Dimensions of candidate evaluation', *The American Political Science Review*, **64**, 1,167–85.

Wheeler, J. O. (1976), 'Location of mobile home manufacturing: a multidimensional scaling analysis', *The Professional Geographer*, **28**, 261–6.

White, H. C., S. A. Boorman and R. L. Breiger (1976), 'Social structure from multiple networks, I: Blockmodels of roles and positions', *American Journal of Sociology*, **81**, 730–80.

Wilson, T. P. (1974), 'Measures of association for bivariate ordinal hypotheses', in H. M. Blalock (ed.) *Measurement in the social sciences*, London: Macmillan.

Wish, M. (1972), 'Differences in perceived similarity of nations', in Romney et al., (eds.) 1974, 290–312.

Wish, M. (1973), *Individual differences in perceptions of dyadic relationships. Notes to accompany tables and figures*, prepared for APA Symposium 'Using multidimensional scaling and related procedures in personality, social, and consumer psychology'. Mimeo.

Wish, M. and J. D. Carroll (1974), 'Applications of individual differences scaling to studies of human perceptions and judgement', in Carterette and Friedman, 1974 (q.v.).

Wishart, D. (1978), CLUSTAN: *User Manual*, 3rd ed., Edinburgh: PLU Report IU/RCS 47.

Wittgenstein, L. (1958), *Philosophical Investigations*, Oxford: Blackwell.

Yoshida, M. (1968), 'Dimensions of tactile impressions', *Japanese Psychological Research*, **10**, 123–37.

Young, F. W. (1968), 'TORSCA-9, A Fortran IV program for nonmetric multidimensional scaling', *Behavioral Science*, **13**, 343–44.

Young, F. W. (1970), 'Nonmetric multidimensional scaling: recovery of metric information', *Psychometrika*, **35**, 455–73.

Young, F. W. (1973), 'The variable monotonic transformation', Psychometric Society Meeting, mimeo.

Young, F. W. (1978), 'A new view of individual differences in the weighted euclidean MDS model', Chapel Hill: University of North Carolina, mimeo.

Young, G. and A. S. Householder (1941), 'Note on multidimensional psychophysical analysis', *Psychometrika*, **6**, 331–3.

Zvulun, E. (1978), 'Multidimensional scalogram analysis', in Shye 1978 (q.v.).

Author Index

Subject Index

point condition
Unfolding model, 139–143,
181–185, 221–225
Unconstrained solutions,
see Internal analysis
Uniform weighting
(rescaling), 235–236

Variance-covariance
matrix, *see*
Dispersion matrix

Vector measures (**A2.1**),
16; *see also* Scalar
products, PM
measures
Vector model, 133–138,
172–176, 187,
208–209, 222–223
properties of, 138
Vector representation of
data, 17, 36–40, 136
Vector weighting models,

211–215

Weak monotonicity, 21, 45–
46, 51–52, 54; *see
also* Ties
Weak order, 5
Weighted unfolding model,
222–223, 225

Young plot, 195–196

Program Index

All programs in Service version 3 (SV3) of the MDS (X) Series are marked by an asterisk (*)
Entries in bold face in brackets immediately following a program name refer to the sections of this
book in which the most crucial aspects of the program are defined.